Experiment for showing by intermittent light the apparently stationary drops into which a fountain is broken up by the action of a musical sound. (*See* page 83).

SOAP-BUBBLES

THEIR COLOURS

AND THE

FORCES WHICH MOLD THEM

*Being the substance of many lectures
delivered to juvenile and popular audiences
with the addition of several new and original sections*

By

C. V. BOYS

With a new preface

By

S. Z. LEWIN

Associate Professor of Chemistry
New York University

DOVER PUBLICATIONS, INC.
NEW YORK

This new Dover edition first published in 1958 is an unabridged and unaltered republication of the last revision, with the addition of a new preface by S. Z. Lewin, Associate Professor of Chemistry, New York University.

International Standard Book Number: 0-486-20542-8
Library of Congress Catalog Card Number: 59-14223

Dover Publications, Inc.
180 Varick Street
New York 14, N. Y.

PREFACE TO DOVER EDITION

EXPERIMENTS WITH SOAP BUBBLES

The experiments and demonstrations described in this charming book by C. V. Boys are not only highly interesting and entertaining; they also illustrate and explain scientific principles to which recent developments have given considerable importance. The general phenomenon of a tension existing in the surface of a drop of liquid, and its manifestations, have applications that range from recent attempts to understand the nature of the atomic nucleus to the theories of the genesis of our planetary system. The rising of liquid in a capillary tube, described on pages 24-27, is still the standard laboratory technique for the precise measurement and comparison of surface tensions. The spreading of oil on water, described on pages 36-39, has been much utilized in studying molecular structure and forces; it is also an important factor in lubrication, in the recovery of oil wells, and in the protection of water reserves against evaporation losses. The concept of surfaces of zero curvature, pages 55-58 and 88-90, and of the minimum surface that can enclose various shaped volumes, pages 120-127, illuminate aspects of the currently very active field in mathematics known as topology. The effect of a magnetic field on oxygen, described on pages 62-64, is now the basis of a widely-used commercial instrument for measuring oxygen concentrations in gases, although a more complicated and expensive (and far less ingenious) set-up than a mere soap bubble is employed to make the effect visible. The very lucid explanation on pages 105-107 of the reason for the very existence of soap bubbles is essentially a statement, in a form that is readily understandable to all, of the fundamental law that is dignified in scientific circles as the "Gibbs Adsorption Isotherm." The solution of soap and glycerine in water that is described in this book for making good soap bubbles is presently widely utilized in the laboratory and in industry to test for gas leaks in valves, pipes and sealed systems. The appearance of bubbles when this solution is painted over a tiny pinhole or crack signals the presence of escaping gas at that place.

In the expectation that many readers of this volume will want to try for themselves the experiments described herein, the following information is offered to facilitate such efforts. Only a soap and glycerine solution is required for the great majority of these experiments. Although the involved directions given by the author for the preparation of a good grade of soap (oleate of soda, pages 171-172) were necessary forty years ago, this is no longer the case. Soaps and detergents are now commercially available in such high quality that they may be used without further purification for these experiments. Any good white soap flakes, such as Lux or Ivory, or synthetic detergents such as Fab, Dreft, or Tide can be recommended. When detergents are used, ordinary tap water can be employed in making up the solutions; otherwise, distilled or de-ionized water is needed.

However, if especially long-lived soap bubbles are desired, the use of extremely pure chemicals is still necessary. A. L. Kuehner has described (in the *Journal of Chemical Education,* vol. 35, page 337, July 1958) a simple procedure for purifying to an adequate degree the oleic acid available from the chemical supply houses, as well as for preparing a dibromo-derivative that yields exceptionally fine bubbles. This latter procedure involves the use of hazardous chemicals (such as liquid bromine) and should not be attempted by an amateur.

An adequately good grade of glycerine (referred to on page 170 as Price's glycerine) may be obtained from any chemical supply house under the designation "Glycerol, U.S.P." (U.S.P. grade chemicals are generally not as pure, and therefore not as expensive as C.P. grade or Reagent grade.)

All the chemicals referred to in this book are available from any chemical supply house (or can be purchased through a local druggist). Two such supply houses are Standard Scientific Co., 30 Turner Place, Piscataway, N. J. 08854 and Fisher Scientific Co., 52 Fadem Road, Springfield, N. J. 07081 (with branches in many other large cities throughout the U.S., Canada, and Europe). In some cases the author's terminology is out-of-date and translation into current chemical nomenclature is necessary. Thus, quicksilver (p. 26) is mercury; benzoline or petrol (p. 35) is obtainable as petroleum ether; bisulfide of carbon (p. 44) is carbon disulfide; "resin" (p. 113) can best be approximated at present by dammar gum or mastic gum;

colophane (p. 113) is colophonium, or rosin; "gutta-percha" (p. 113) is rubber latex; and oil of bitter almonds (p. 158) is benzaldehyde. Instead of a "clock shade" (p. 44), one can use small battery jars, obtainable from the laboratory supply house; a "tap" (p. 44, 50) is a stopcock or valve.

Finally, the reader should be cautioned against using unfamiliar chemicals with anything but the greatest of care. The vapors of many are poisonous; e.g., mercury, aniline, benzene and toluidine (which are mentioned in connection with several of the experiments). Although these are much less volatile than ether or carbon disulfide, they do vaporize enough to be dangerous, and they should never be worked with in an enclosed place. No traces of these substances should be left around on tables or floors exposed to the air after experimentation is complete. They will continue to contribute vapors to the air that is breathed and will act as cumulative poisons in the body. If liquid chemicals are spilled on the skin they must be washed off immediately with a copious flow of water, and this should be followed by scrubbing with soap and water.

New York   S. Z. LEWIN,
June, 1958   Associate Professor of Chemistry,
                            New York University

PREFACE

I WOULD ask those readers who have grown up, and who may be disposed to find fault with this book, on the ground that in so many points it is incomplete, or that much is so elementary or well known, to remember that the lectures were meant for juveniles, and for juveniles only. These latter I would urge to do their best to repeat the experiments described. They will find that in many cases no apparatus beyond a few pieces of glass or india-rubber pipe, or other simple things easily obtained are required. If they will take this trouble they will find themselves well repaid, and if instead of being discouraged by a few failures they will persevere with the best means at their disposal, they will soon find more to interest them in experiments in which they only succeed after a little trouble than in those which go all right at once. Some are so simple that no help can be wanted, while some will probably be too difficult, even with assistance ; but to encourage those who wish to see for themselves the experiments that I have described, I have given such hints at the end of the book as I thought would be most useful.

I have freely made use of the published work of many distinguished men, among whom I may mention Savart, Plateau, Clerk Maxwell, Sir William Thomson, Lord Rayleigh, Mr. Chichester Bell, and Prof. Rücker. The experiments have mostly been described by them, some have been taken from journals, and I have devised or arranged a few. I am also indebted to Prof. Rücker for the apparatus illustrated in Figs. 23, 24, 26, 27, 30, 31 and 32.

PREFACE TO THE NEW AND ENLARGED EDITION

As the earlier editions of this book have met with so favourable a reception, since in fact about two tons of my bubbles are floating about the world, and the book has been translated into French, German and Polish, I have thought fit to rearrange, alter and enlarge it. The chapter on the colours and thicknesses of bubbles is entirely new, as are two or three other shorter ones on bubbles of different kinds. In some of these, especially that on their colours, the treatment of the subject is necessarily a good deal more difficult than it is in the original parts. As the book is primarily intended for the general reader rather than for the student of physical science I have avoided the use of all trigonometrical and algebraical formulæ, as I know their paralysing effect on the non-technical reader. At the same time I do not think that there is any want of precision or accuracy as a result. I have therefore been compelled to employ a more cumbersome arithmetical treatment in some cases, while in others I have used geometrical construction in order to obtain quantitative results. This has the advantage of providing ocular demonstration as well as proof, and in the case of the loss of time within the thickness of a soap film is far neater and more natural than the more usual trigonometrical method. I have felt constrained to use the archaic British units of measurement, as the unfamiliar metric terminology would have distracted the attention of the majority for whom this book is intended, who have spent untold hours that might have gone into mathematical or general education in performing ridiculous operations such as reduction, compound multiplication and practice which our British methods of measurement necessitate, but which in more enlightened countries are wholly unnecessary. This book is not prepared to meet the requirements and artificial restrictions of any syllabus, and it is not prepared to help students through any examination.   I cannot help thinking, however, that if the type of student who puts more faith in learning formulæ than in understanding how they may be recovered when forgotten, as they will be, would condescend to spend the time necessary for reading the chapter on the colours of soap-bubbles he would derive some help from it, and he might even find it useful in preparing for an examination.

In the additional chapters I have found it more convenient to give with the text sufficient guidance for the repetition of experiments instead of reserving this for the practical hints at the end, which remain much as they were.

December, 1911.

CONTENTS

SOAP-BUBBLES

THEIR COLOURS

AND THE

FORCES WHICH MOULD THEM

Introductory

I DO not suppose that there is any one in this room who has not occasionally blown a common soap-bubble, and while admiring the perfection of its form, and the marvellous brilliancy of its colours, wondered how it is that such a magnificent object can be so easily produced.

I hope that none of you are yet tired of playing with bubbles, because, as I hope we shall see, there is more in a common bubble than those who have only played with them generally imagine.

The wonder and admiration so beautifully portrayed by Millais in a picture, copies of which, thanks to modern advertising enterprise, some of you may possibly have seen, will, I hope, in no way fall away in consequence of these lectures ; I think you will find that it will grow as your knowledge of the subject increases. Plateau in his famous work, *Statique des Liquides*, quotes a passage from a book by Henry Berthoud, to the effect that there is an Etruscan vase in the Louvre in Paris in which children are represented blowing bubbles from a pipe. Plateau states, however, that no classical author refers to any such amusement, and the only two references to bubbles of any kind that he can find are in Ovid and

Martial. I have hunted for this vase at the Louvre in vain. A correspondent, however, sent the quotation to the director, by whom he was informed that no such vase was there, but that a number of fictitious antique vases had been removed from the collection.

It is possible that some of you may like to know why I have chosen soap-bubbles as my subject; if so, I am glad to tell you. Though there are many subjects which might seem to a beginner to be more wonderful, more brilliant, or more exciting, there are few which so directly bear upon the things which we see every day. You cannot pour water from a jug or tea from a tea-pot; you cannot even do anything with a liquid of any kind, without setting in action the forces to which I am about to direct your attention. You cannot then fail frequently to be reminded of what you will hear and see in this room, and, what is perhaps most important of all, many of the things I am going to show you are so simple that you will be able without any apparatus to repeat for yourselves the experiments which I have prepared, and this you will find more interesting and instructive than merely listening to me and watching what I do.

There is one more thing I should like to explain, and that is why I am going to show experiments at all. You will at once answer, because it would be so dreadfully dull if I didn't. Perhaps it would. But that is not the only reason. I would remind you then that when we want to find out anything that we do not know, there are two ways of proceeding. We may either ask somebody else who does know, or read what the most learned men have written about it, which is a very good plan if anybody happens to be able to answer our question; or else we may adopt the other plan, and by arranging an experiment, find out for ourselves. An experiment is a question which we ask of Nature, who is always ready to give a correct answer, provided we ask properly, that is, provided we arrange a proper experiment. An

experiment is not a conjuring trick, something simply to
make you wonder, nor is it simply shown because it is
beautiful, or because it serves to relieve the monotony of
a lecture ; if any of the experiments I show are beautiful,
or do serve to make these lectures a little less dull, so much
the better ; but their chief object is to enable you to see
for yourselves what the true answers are to the questions
that I shall ask.

The Elastic Skin of Liquids

Now I shall begin by performing an experiment which
you have all probably tried dozens of times without
recognizing that you were making an experiment at all.
I have in my hand a common camel's-hair brush.   If you
want to make the hairs cling together and come to a
point, you wet it, and then you say the hairs cling together because the brush is wet.   Now let us try the
experiment ; but as you cannot see this brush across the
room, I hold it in the lantern, and you can see it enlarged upon the screen (Fig. 1, left hand).   Now it is
dry, and the hairs are separately visible.   I am now
dipping it in the water, as you can see, and on taking it
out, the hairs, as we expected, cling together (Fig. 1,
right hand), because they are wet, as we are in the habit
of saying.   I shall now hold the brush in the water, but
there it is evident that the hairs do not cling at all (Fig.
1, middle), and yet they surely are wet now, being actually
in the water.   It would appear then that the reason
which we always give is not exactly correct.   This
experiment, which requires nothing more than a brush
and a glass of water, then, shows that the hairs of a brush
cling together not only because they are wet, but for
some other reason as well which we do not yet know.   It
also shows that a very common belief as to opening our
eyes under water is not founded on fact.   It is very

commonly said that if you dive into the water with your
eyes shut you cannot see properly when you open them
under water, because the water gums the eyelashes down
over the eyes ; and therefore you must dive in with your
eyes open if you wish to see under water. Now as a
matter of fact this is not the case at all ; it makes no

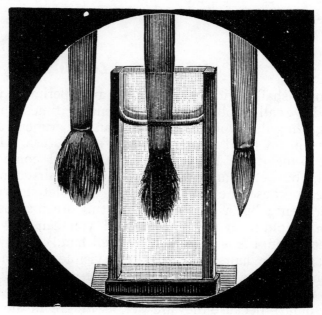

FIG. 1.

difference whether your eyes are open or not when you
dive in, you can open them and see just as well either
way. In the case of the brush we have seen that water
does not cause the hairs to cling together or to anything
else when under the water, it is only when taken out that
this is the case. This experiment, though it has not
explained why the hairs cling together, has at any rate
told us that the reason always given is not sufficient.

I shall now try another experiment as simple as the

last. I have a pipe from which water is very slowly issuing, but it does not fall away continuously ; a drop forms which slowly grows until it has attained a certain definite size, and then it suddenly falls away. I want you to notice that every time this happens the drop is always exactly the same size and shape. Now this cannot be mere chance ; there must be some reason for the definite

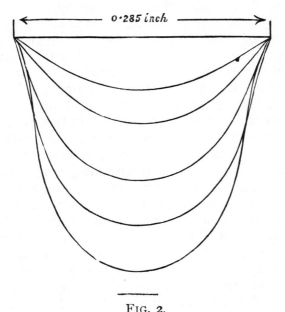

FIG. 2.

size and shape. Why does the water remain at all? It is heavy and is ready to fall, but it does not fall; it remains clinging until it is a certain size, and then it suddenly breaks away, as if whatever held it was not strong enough to carry a greater weight. Mr. Worthington has carefully drawn on a magnified scale the exact shape of a drop of water of different sizes, and these you now see upon the diagram on the wall (Fig. 2). These diagrams will probably suggest the idea that the water is hanging suspended in an elastic bag, and that the bag

B

breaks or is torn away when there is too great a weight
for it to carry.   It is true there is no bag at all really, but
yet the drops take a shape which suggests an elastic bag.
To show you that this is no fancy, I have supported by
a tripod a large ring of wood over which a thin sheet of
india-rubber has been stretched, and now on allowing
water to pour in from this pipe you will see the rubber
slowly stretching under the increasing weight, and, what
I especially want you to notice, it always assumes a form
like those on the diagram.   As the weight of water
increases the bag stretches, and now that there is about a
pailful of water in it, it is getting to a state which
indicates that it cannot last much longer; it is like the
water-drop just before it falls away, and now suddenly it
changes its shape (Fig. 3), and it would immediately tear
itself away if it were not for the fact that india-rubber
does not stretch indefinitely; after a time it gets tight and
will withstand a greater pull without giving way.   You
therefore see the great drop now permanently hanging
which is almost exactly the same in shape as the water-
drop at the point of rupture.   I shall now let the water
run out by means of a syphon, and then the drop slowly
contracts again.   Now in this case we clearly have a
heavy liquid in an elastic bag, whereas in the drop of
water we have the same liquid but no bag that is visible.
As the two drops behave in almost exactly the same way,
we should naturally be led to expect that their form and
movements are due to the same cause, and that the
small water-drop has something holding it together like
the india-rubber you now see.

   Let us see how this fits the first experiment with the
brush.   That showed that the hairs do not cling together
simply because they are wet; it is necessary also that
the brush should be taken out of the water, or in other
words it is necessary that the surface or the skin of the
water should be present to bind the hairs together.   If
then we suppose that the surface of water is like an

elastic skin, then both the experiments with the wet brush and with the water-drop will be explained.

Let us therefore try another experiment to see

FIG. 3.

whether in other ways water behaves as if it had an elastic skin.

I have here a plain wire frame fixed to a stem with a weight at the bottom, and a hollow glass globe fastened

B 2

to it with sealing-wax. The globe is large enough to make the whole thing float in water with the frame up in the air. I can of course press it down so that the frame touches the water. To make the movement of the frame more evident there is fixed to it a paper flag.

Now if water behaves as if the surface were an elastic skin, then it should resist the upward passage of the

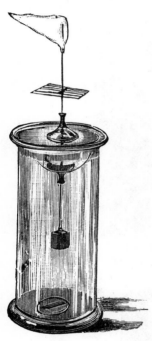

frame which I am now holding below the surface. I let go, and instead of bobbing up as it would do if there were no such action, it remains tethered down by this skin of the water. If I disturb the water so as to let the frame out at one corner, then, as you see, it dances up immediately (Fig. 4). You can see that the skin of the water must have been fairly strong, because a weight of about one quarter of an ounce placed upon the frame is only just sufficient to make the whole thing sink.

This apparatus, which was originally described by Van der Mensbrugghe, I shall make use of again in a few minutes.

I can show you in a more striking way that there is this elastic layer or skin on pure clean water. I have a small

FIG. 4.

sieve made of wire gauze sufficiently coarse to allow a common pin to be put through any of the holes. There are moreover about eleven thousand of these holes in the bottom of the sieve. Now, as you know, clean wire is wetted by water, that is, if it is dipped in water it comes out wet; on the other hand, some materials, such as paraffin wax, of which paraffin candles are made, are not wetted or really touched by water, as you may see for yourselves if you will only dip a paraffin candle into water. I have melted a quantity of paraffin in a dish and dipped this gauze into the melted paraffin so as to coat the wire all over with it, but I have shaken it well while hot to knock the paraffin out of the holes. You can now see on the screen that the holes, all except one or two, are open, and that a common pin can be passed through readily enough. This then is the apparatus. Now if water has an elastic skin which it requires force to stretch, it ought not to run through these holes very readily; it ought not to be able to get through at all unless forced, because at each hole the skin would have to be stretched to allow the water to. get to the other side. This you understand is only true if the water does nót wet or really touch the wire. Now to

FIG. 5.

prevent the water that I am going to pour in from striking the bottom with so much force as to drive it through, I have laid a small piece of paper in the sieve, and am pouring the water on to the paper, which breaks the fall (Fig. 5). I have now poured in about half a tumbler of water, and I might put in more. I take away the paper but not a drop runs through. If I give the sieve a jolt then the water is driven to the other side, and in a moment it has all escaped. Perhaps this will remind you of one of the exploits of our old friend Simple Simon,

"Who went for water in a sieve,
But soon it all ran through."

But you see if you only manage the sieve properly, this is not quite so absurd as people generally suppose.

If now I shake the water off the sieve, I can, for the

same reason, set it to float on water, because its weight is not sufficient to stretch the skin of the water through all the holes. The water, therefore, remains on the other side, and it floats even though, as I have already said, there are eleven thousand holes in the bottom, any one of which is large enough to allow an ordinary pin to pass through. This experiment also illustrates how difficult it is to write real and perfect nonsense.

You may remember one of the stories in Lear's book of Nonsense Songs.

Fɪɢ. 6.

" They went to sea in a sieve, they did,
In a sieve they went to sea :
In spite of all their friends could say,
On a winter's morn, on a stormy day,
In a sieve they went to sea.
*   *   *   *
" They sailed away in a sieve, they did,
In a sieve they sailed so fast,
With only a beautiful pea-green veil,
Tied with a riband by way of a sail,
To a small tobacco-pipe mast ; "
*   *   *   *

And so on. You see that it is quite possible to go to sea in a sieve—that is, if the sieve is large enough and the water is not too rough—and that the above lines are now realized in every particular (Fig. 6).

I may give one more example of the power of this elastic skin of water. If you wish to pour water from a tumbler into a narrow-necked bottle, you know how if you pour slowly it nearly all runs down the side of the glass and gets spilled about, whereas if you pour quickly there is no room for the great quantity of water to pass into the bottle all at once, and so it gets spilled again. But if you take a piece of stick or a glass rod, and hold it against the edge of the tumbler, then the water runs down the rod and into the bottle, and none is lost (Fig. 7); you may even hold the rod inclined to one side, as I am now doing, but the water runs down the wet rod because this elastic skin forms a kind of tube which prevents the water from escaping. This action is often made use of in the country to carry the water from the gutters under the roof into a water-butt below.

FIG. 7.

A piece of stick does nearly as well as an iron pipe, and it does not cost anything like so much.

I think then that I have now done enough to show that on the surface of water there is a kind of elastic skin. I do not mean that there is anything that is not water on the surface, but that the water while there acts in a different way to what it does inside, and that it acts as if it were an elastic skin made of something like very thin india-rubber, only that it is perfectly and absolutely elastic, which india-rubber is not.

Capillary Attraction

You will now be in a position to understand how it is that in narrow tubes water does not find its own level, but behaves in an unexpected manner. I have placed in front of the lantern a dish of water coloured blue so that you may the more easily see it. I shall now dip into the water a very narrow glass pipe, and immediately the water rushes up and stands about half an inch above the general level. The tube inside is wet. The elastic skin of the water is therefore attached to the tube, and goes on pulling up the water until the weight of the water raised above the general level is equal to the force exerted by the skin. If I take a tube about twice as big, then this pulling action which is going on all round the tube will cause it to lift twice the weight of water, but this will not make the water rise twice as high, because the larger tube holds so much more water for a given length than the smaller tube. It will not even pull it up as high as it did in the case of the smaller tube, because if it were pulled up as high the weight of the water raised would in that case be four times as great, and not only twice as great, as you might at first think. It will therefore only raise the water in the larger tube to half the height, and now that the two tubes are side by side you see the water in the smaller tube standing twice as high as it does in the larger tube. In the same way, if I were to take a tube as fine as a hair the water would go up ever so much higher. It is for this reason that this is called Capillarity, from the Latin word *capillus*, a hair, because the action is so marked in a tube the size of a hair.

Supposing now you had a great number of tubes of all sizes, and placed them in a row with the smallest on one side and all the others in the order of their sizes, then it is evident that the water would rise highest in the smallest tube and less and less high in each tube in the

row (Fig. 8), until when you came to a very large tube
you would not be able to see that the water was raised at
all.   You can very easily obtain the same kind of effect
by simply taking two squares pieces of window glass and
placing them face to face with a common match or small
fragment of anything to keep them a small distance apart
along one edge while they meet together along the

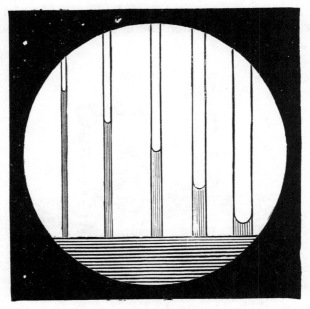

FIG. 8.

opposite edge.   An india-rubber ring stretched over
them will hold them in this position.   I now take such
a pair of plates and stand it in a dish of coloured water,
and you at once see that the water creeps up to the top
of the plates on the edge where they meet, and as the
distance between the plates gradually increases, so the
height to which the water rises gradually gets less, and
the result is that the surface of the liquid forms a

beautifully regular curve which is called by mathematicians a rectangular hyperbola (Fig. 9). I shall have presently to say more about this and some other curves, and so I shall not do more now than state that the hyperbola is formed because as the width between the plates gets greater the height gets less, or, what comes to the same thing, because the weight of liquid supported at any small part of the curve is always the same.

If the plates or the tubes had been made of material not wetted by water, then the effect of the tension of the

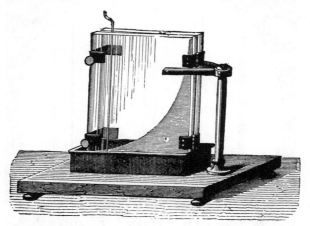

Fig. 9.

surface would be to drag the liquid away from the narrow spaces, and the more so as the spaces were narrower. As it is not easy to show this well with paraffined glass plates or tubes and water, I shall use another liquid which does not wet or touch clean glass, namely, quicksilver. As it is not possible to see through quicksilver, it will not do to put a narrow tube into this liquid to show that the level is lower in the tube than in the surrounding vessel, but the same result may be obtained by having a wide and a narrow tube joined together. Then, as you see upon the screen, the quicksilver is lower in the narrow than in the wide tube, whereas in a similar apparatus the reverse is the case with water (Fig. 10).

Although the elastic tension which I have called the strength of the water-skin is very small where big things are considered, this is not the case where very small things are acted upon. For instance, those of you who are fortunate enough to live in the country and who have gone down to play by the side of a brook must often have seen water-spiders and other small creatures running on the water without sinking in. For some reason their

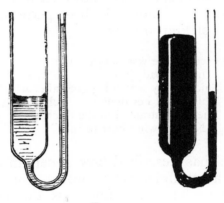

FIG. 10.

feet are not wetted by the water, and so they tread down and form a small dimple where each foot rests, and so the up-pulling sides of the dimples just support the weight of the creature. It follows also from this that this weight is exactly equal to the weight of the water that would just fill all the dimples up to the general water level, that is, supposing that it were possible to imagine the dimpled water solidified for the purpose of the experiment. Mr. H. H. Dixon, of Dublin, once very ingeniously measured the force with which one of these water-spiders pressed its different feet while running on the water. He photographed the shadow of the spider and of the dimples upon a white porcelain dish containing the water on which the spider ran. He then mounted one of the spider's feet on a very delicate balance, and made it press on the water with different degrees of pressure, and again photographed the shadow of the dimple for each degree of pressure. He was thus able to make a scale by means of which he could tell the pressure exerted by the spider on any foot by comparing the size of the shadow of the dimples with those on his scale. He was also able to see the order in which the water-spider put down its feet, and so solve the problem for the spider which so perplexed the centipede of the well-known lines—

> "A centipede was happy quite,
>   Until a toad, in fun,
>   Said, 'Pray which leg goes after which?'
>   This raised her doubts to such a pitch
>   She fell distracted in the ditch,
>   Not knowing how to run."

Prof. Miall has described how a certain other water-spider spins a net under water through which air will not pass, just as water would not pass through the sieve which it did not wet. The spider then goes to the surface and carries air down and liberates it under the net and so gradually accumulates a reservoir of air to enable it to breathe at leisure when it has a good supply.

The elastic water-skin is made use of by certain larvæ which live immersed in the water, as well as by water-spiders and other creatures which run over it. The common gnat lays its eggs in stagnant water, and seems especially fond of water-butts and troughs in gardens or greenhouses. These eggs in time hatch out into the larvæ of the gnat which corresponds to the silk-worm or caterpillar of a moth or butterfly. This larva may be found in thousands in hot weather in ordinary rainwater-butts. You must have noticed small, dark

creatures which swim in a curious jerky way and go to
the bottom when you suddenly come near and frighten
them. However, if you keep still, you will not have long
to wait before you see them swimming back again to the
surface, to which they attach themselves, and then remain
hanging. It is very easy to show these alive on the
screen. In the place of a lantern slide I have a cell
containing water in which are a number of these larvæ.
You will see how they swim to the surface and then hang

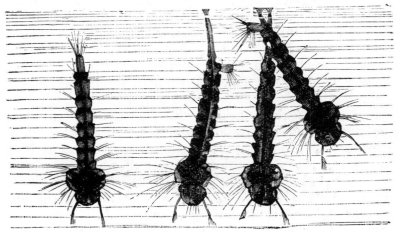

FIG. II.

by a projection like a tail (Fig. 11). This is a breathing-
tube, and so, even though they are heavier than water
and naturally sink, they are able to hang by their breath-
ing-tubes, and breathe while they can eat rotten leaves
without exertion. The larva on the left side of the
figure had broken away from the surface just before I
took the photograph and was slowly sinking.

If you examine the surface of water, say in a tumbler
in which you have placed some of these larvæ, you will
see, when they come to the surface, as they will have to
do, that there is a small dimple where they hang, and

the weight of the water that would be needed to fill this dimple is exactly equal to the downward pull of the larva. If you look through a magnifying-glass you will see it much better and also notice what an interesting looking creature the larva really is.

Prof. Miall has also shown us how the common duckweed makes use of surface tension to turn itself round so that, if the pond is not already crammed with as many as it will hold, the leaves attract each other so as to rest in contact end to end, leaving their sides, from which the young plants originate by a process of budding, free for their development. The leaf has a central ridge, and the heel and toe, so to speak, are above the general surface, so that the water surface is curved up to these. Now it is a fact that things that are wetted by water so that the water is curved upwards where it meets them, attract one another, the reason being that the pressure within the raised water is less than the atmospheric pressure on the other side in proportion as it is raised higher. Also things neither of which is wetted by water attract each other, as in that case the water is curved downwards where it meets them, and the air-pressure in the region of the depression is less than that of the water on the other side in proportion as it is depressed lower. On the other hand, two things, one of which is wetted and one of which is not wetted, repel each other. Now, coming back to the duckweed, the elevated toe and heel attract each other as wetted things do, while the sides, being depressed to the water level about, are neutral. You will do well to make the experiment yourself. Duckweed is easily found in the country. Fill a tumbler until it is rather over full, *i. e.* so that the curvature of the water at the edge is downwards. Small things that are wetted will then remain in the middle. Place a few duckweed plants in the water and see how they attract each other endways only, and how a wetted point will cause isolated ones to turn round almost like a magnet acting on a compass needle. Having got your duckweed, it will be found interesting to keep it a few days and watch the budding and separation of new plants.

So far I have given you no idea what force is exerted by this elastic skin of water. Measurements made with narrow tubes, with drops, and in other ways, all show that it is almost exactly equal to the weight of three and a quarter grains to the inch. We have, moreover, not yet seen whether other liquids act in the same way, and if so whether in other cases the strength of the elastic skin is the same.

Capillarity of Different Liquids

You now see a second tube identical with that from which drops of water were formed, but in this case the liquid is alcohol. Now that drops are forming, you see

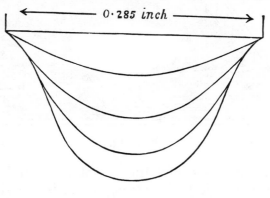

FIG. 12.

at once that while alcohol makes drops which have a definite size and shape when they fall away, the alcohol drops are not by any means so large as the drops of water which are falling by their side. Two possible reasons might be given to explain this. Either alcohol is a heavier liquid than water, which would account for the smaller drop if the skin in each liquid had the same strength, or else if alcohol is not heavier than water its skin must be weaker than the skin of water. As a matter of fact alcohol is a lighter liquid than water, and so still more must the skin of alcohol be weaker than that of water.

We can easily put this to the test of experiment. In

FIG. 13.

the game that is called the tug-of-war you know well enough which side is the strongest; it is the side which pulls the other over the line. Let us then make alcohol and water play the same game. In order that you may see the water, it is coloured blue. It is lying as a shallow layer on the bottom of this white dish. At the present time the skin of the water is pulling equally in all directions, and so nothing happens; but if I pour a few drops of alcohol into the middle, then at the line

which separates the alcohol from the water we have
alcohol on one side pulling in, while we have water on
the other side pulling out, and you see the result. The
water is victorious; it rushes away in all directions,
carrying a quantity of the alcohol away with it, and leaves
the bottom of the dish dry (Fig. 13).

This difference in the strength of the skin of alcohol

FIG. 14.

and of water, or of water containing much or little alcohol,
gives rise to a curious motion which you may see on the
side of a wine-glass in which there is some fairly strong
wine, such as port. The liquid is observed to climb up
the sides of the glass, then to gather into drops, and to
run down again, and this goes on for a long time. This
was explained by Professor James Thomson as follows:
The thin layer of wine on the side of the glass being
exposed to the air, loses its alcohol by evaporation more

quickly than the wine in the glass does. It therefore becomes weaker in alcohol or stronger in water than that below, and for this reason it has a stronger skin. It therefore pulls up more wine from below, and this goes on until there is so much that drops form, and it runs back again into the glass, as you now see upon the screen (Fig. 14). It is probable that this movement is referred to in Proverbs xxiii. 31 : " Look not thou upon the wine when it is red, when it giveth his colour in the cup, when it moveth itself aright."

Ether, in the same way, has a skin which is weaker than the skin of water. The very smallest quantity of ether on the surface of water will produce a perceptible effect. For instance, the wire frame which I left some time ago is still resting against the water-skin. The buoyancy of the glass bulb is trying to push it through, but the upward force is just not sufficient. I will however pour a few drops of ether into a glass, and simply pour the vapour upon the surface of the water (not a drop of *liquid* is passing over), and almost immediately sufficient ether has condensed upon the water to reduce the strength of the skin to such an extent that the frame jumps up out of the water.

There is a well-known case in which the difference between the strength of the skins of two liquids may be either a source of vexation or, if we know how to make use of it, an advantage. If you spill grease on your coat you can take it out very well with benzene. Now if you apply benzene to the grease, and then apply fresh benzene to that already there, you have this result—there is then greasy benzene on the coat to which you apply fresh benzene. It so happens that greasy benzene has a stronger skin than pure benzene. The greasy benzene therefore plays at tug-of-war with pure benzene, and being stronger wins and runs away in all directions, and the more you apply benzene the more the greasy benzene runs away carrying the grease with it. But if you follow the proper method, and first make a ring of clean benzene round the grease-spot, and then apply benzene to the grease, you then have the greasy benzene running away from the pure benzene ring and heaping itself together in the middle, and escaping into the fresh rag that you apply, so that the grease is all of it removed.

I put this to a very severe test once when a new white satin dress had been spoiled by an upset of soup. I laid a number of dusters on the ground out of doors and carefully laid the part which was stained over the dusters, and then using a quart or so of pure benzene (not benzoline or petrol which by comparison are useless) I poured it freely first in a ring round, and then through the place constantly, replacing the old dusters by new. Then on lifting it up the remaining benzene quickly evaporated, and **no trace** of the stain nor any " high-water mark " could be detected.

There is a difference again between hot and cold grease, as you may see, when you get home, if you watch a common candle burning. Close to the flame the grease is hotter than it is near the outside. It has there-fore a weaker skin, and so a perpetual circulation is kept up, and the grease runs out on the surface and back again below, carrying little specks of dust which make this movement visible, and making the candle burn regularly.

You probably know how to take out grease-stains with a hot poker and blotting-paper. Here again the same kind of action is going on.

A piece of lighted camphor floating on water is another example of movement set up by differences in the strength of the skin of water owing to the action of the camphor.

The best way to make the experiment with camphor is to take a large basin of very clean water and then holding the camphor over the water, scrape the corners lightly with a knife. The minute specks of camphor which fall on the water will then display an activity which is quite surprising. The water however must be nearly perfectly free from grease, mere contact with the finger for a moment may be sufficient to stop the display of activity. Lord Rayleigh has determined the weight of oil that will "kill" camphor in water in a large bath, and has found in this way that if the thickness of oil is no more than $\frac{1}{12000000}$ inch, the camphor is dead. If no vessel in the house is clean enough, a rain-water butt which has been overflowing during the rain will be found perfectly satisfactory. It would be best not to tell the cook that all the clean things are greasy.

The camphor experiment is sufficient to show that a tiny speck of oil spreads almost instantly over a large surface of water. The strength of the pure water skin is greater than that of the two separate skins, one where the oil and air meet, and the other where the oil and the water meet, and thus it is that the oil spreads first so as to show colours like a soap-bubble, and then so much thinner that no colours can be seen.

I am not sure whether the extremely unpleasant and clinging taste of castor oil is not due to an action of this kind causing the oil to spread over every part of the mouth, and whether the use of ginger-wine in making the oil less unpleasant may not depend upon the reduction of the strength of the water skin in the mouth by the action of the alcohol in the wine so that the oil no longer searches out every corner and crevice, but goes in the required direction like an oyster. Of course the hot flavour is all to the good, but that alone would seem hardly sufficient.

Pouring Oil on Troubled Water

Oil besides spreading on water has a wonderful effect in preventing the formation of ripples by the wind, for when a wave of any size whatever is moving in the water the sloping side of the wave in front of the crest is being made less in extent while that behind is being stretched. Now the thin film of oil where the surface is becoming greater will become thinner and the strength of the water skin there will increase, while where the surface is becoming less the oil will thicken and the strength will become less. On each side of the wave therefore forces are set up opposing the existence of the wave, and though these are immaterial in the case of long waves they are most potent where the waves are very small. Thus it is that the roughening of the water by the wind and the hold of the wind on the water is so greatly affected by a film of oil. If any of you should be living on a yacht you will be surprised how long you can see the white smooth place where you have thrown the sardine oil overboard.

Fig. 15 is a photograph of part of the Serpentine in Hyde Park opposite the gun-powder magazine, taken on a very windy day. A spoonful of olive oil had been thrown into the water. A large oily tract from twenty to thirty yards in diameter is clearly seen, showing how the thin layer of oil prevented the ripples from forming. As the smooth patch of water would be about 1000 times as long and 1000 times as wide as the spoon, the depth of the oil would be about $\frac{1}{1000000}$th of the depth of the oil in the spoon, or perhaps $\frac{1}{10000000}$th of an inch deep. The effect of pouring oil on the troubled waters of the peaceful Serpentine is evident.

I have no doubt that if the morning bathers were to look at the water to leeward on a windy day they would see that they had had the same effect as the oil. It is for the same reason that the track of steamships at sea is visible as a smooth lane for so long.

I will give only one more example.

If you are painting in water-colours on greasy paper or certain shiny surfaces the paint will not lie smoothly on the paper, but runs together in the well-known way; a very little ox-gall, however, makes it lie perfectly, because ox-gall so reduces the strength of the skin of water that

FIG. 15.

it will wet surfaces that pure water will not wet. This reduction of the surface tension you can see if I use the same wire frame a third time. The ether has now evaporated, and I can again make it rest against the surface of the water, but very soon after I touch the water with a brush containing ox-gall the frame jumps up as suddenly as before.

The reduction of the surface tension of water by ox-gall or by soap may be made apparent if you have some gnat larvæ in a tumbler, for as soon as either of these touch the surface of the clean water the creatures are no longer able to support themselves by their breathing tubes. Oil or petroleum dropped on the water in water-butts and pools will destroy the larvæ of the gnat and mosquito, and this treatment is now largely practised with a view of reducing malaria and other diseases which are carried by the mosquito.

It is quite unnecessary that I should any further insist upon the fact that the outside of a liquid acts as if it were a perfectly elastic skin stretched with a certain definite force.

Liquid Drops

Suppose now that you take a small quantity of water, say as much as would go into a nutshell, and suddenly let it go, what will happen? Of course it will fall down and be dashed against the ground. Or again, suppose you take the same quantity of water and lay it carefully upon a cake of paraffin wax dusted over with lycopodium which it does not wet, what will happen? Here again the weight of the drop—that which makes it fall if not held—will squeeze it against the paraffin and make it spread out into a flat cake. What would happen if the weight of the drop or the force pulling it downwards could be prevented from acting? In such a case the drop would only feel the effect of the elastic skin; which would try to pull it into such a form as to make the surface as small as possible. It would in fact rapidly become a perfectly round ball, because in no other way can so small a surface be obtained. If, instead of taking so much water, we were to take a drop about as large as a pin's head, then the weight which tends to squeeze it out or make it fall would be far less, while the skin would be just as strong, and would in reality have a greater moulding power, though why I cannot now explain. We should therefore expect that by taking a sufficiently small quantity of water the moulding power of the skin would ultimately be able almost entirely to counteract

FIG. 16.

the weight of the drop, so that very small drops should appear like perfect little balls. If you have found any difficulty in following this argument, a very simple illustration will make it clear. You many of you probably know how by folding paper to make this little thing which I hold in my hand (Fig. 16). It is called a cat-box, because of its power of dispelling cats when it is filled with water and well thrown. This one, large enough to hold about half a pint, is made out of a small piece of the *Times* newspaper. You may fill it with water and carry it about and throw it with your full power, and the strength of the paper skin is sufficient to hold it together until it hits anything, when of course it bursts and the water comes out. On the other hand, the large one made out of a whole sheet of the *Times* is barely able to withstand the weight of the water that it will hold. It is only just strong enough to allow of its being filled and carried, and then it may be dropped from a height, but you cannot throw it. In the same way the weaker

skin of a liquid will not make a large quantity take the
shape of a ball, but it will mould a minute drop so
perfectly that you cannot tell by looking at it that it is
not perfectly round every way. This is most easily seen
with quicksilver. A large quantity rolls about like a flat
cake, but the very small drops obtained by throwing

FIG. 17.

some violently on the table and so breaking it up appear
perfectly round. You can see the same difference in the
beads of gold now upon the screen (Fig. 17). They are
now solid, but they were melted and then allowed to cool
without being disturbed. Though the large bead is
flattened by its weight, the small one appears perfectly
round. Finally, you may see the same thing with water
if you dust a little lycopodium on the table. Then
water falling will roll itself up into perfect little balls.

You may even see the same thing on a dusty day if you water the road with a watering-pot.

If it were not for the weight of liquids, that is the force with which they are pulled down towards the earth, large drops would be as perfectly round as small ones. This was first beautifully shown by Plateau, the blind experimentalist, who placed one liquid inside another which is equally heavy, and with which it does not mix. Alcohol is lighter than oil, while water is heavier, but a suitable mixture of alcohol and water is just as heavy as oil, and so oil does not either tend to rise or to fall when im-

Fig. 18.

mersed in such a mixture. I have in front of the lantern a glass box containing alcohol and water, and by means of a tube I shall slowly allow oil to flow in. You see that as I remove the tube it becomes a perfect ball an inch or more in diameter (Fig. 18). There are now two or three of these balls of oil all perfectly round. I want you to notice that when I hit them on one side the large balls recover their shape slowly, while the small ones become round again much more quickly. There is a very beautiful effect which can be produced with this apparatus, and though it is not necessary to refer to it, it is well worth while now that the apparatus is set up to show it to you. In the middle of the box there is an axle with a disc upon it to which I can make the oil adhere. Now if I

slowly turn the wire and disc the oil will turn also. As
I gradually increase the speed the oil tends to fly away
in all directions, but the elastic skin retains it. The
result is that the ball becomes flattened at its poles like
the earth itself. On increasing the speed, the tendency
of the oil to get away is at last too much for the elastic
skin, and a ring breaks away (Fig. 19), which almost
immediately contracts again on to the rest of the ball as
the speed falls. If I turn it sufficiently fast the ring breaks
up into a series of balls which you now see. One cannot

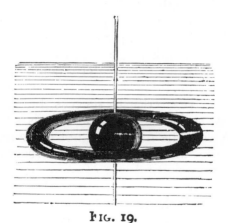

FIG. 19.

help being reminded of the heavenly bodies by this beautiful experiment of Plateau's, for you see a central body and
a series of balls of different sizes all turning and travelling
round in the same direction; but the forces which are
acting in the two cases are totally distinct, and what you
see has nothing whatever to do with the sun and the
planets.

We have thus seen that a large ball of liquid can be
moulded by the elasticity of its skin if the disturbing
effect of its weight is neutralized, as in the last experiment. This disturbing effect is practically of no account
in the case of a soap-bubble, because it is so thin that it

hardly weighs anything. You all know, of course, that a soap-bubble is perfectly round, and now you know why; it is because the elastic film, trying to become as small as it can, must take the form which has the smallest surface for its content, and that form is the sphere. I want you to notice here, as with the oil, that a large bubble oscillates much more slowly than a small one when knocked out of shape with a bat covered with baize or wool.

It is rather difficult to make the experiment with water, alcohol and oil mainly because if the oil has the same density as the surrounding liquid at one temperature it becomes lighter if the temperature rises, or heavier if it falls. The oil expands more than the mixture with rise of temperature, hence the greater change of density. In a later part of this book I refer to another mixture of liquids that I have used with success, but the bisulphide of carbon has so bad a smell, and it is so dangerously inflammable, that the mixture is not to be recommended for general use. Mr. C. R. Darling has recently described a very easy and beautiful way of showing a large liquid sphere. The best vessel to use that can easily be obtained is a clock shade with flat sides so as to avoid the magnification sideways by the curved sides of a round vessel. Make a solution of three parts by weight of common salt in 100 of water, but do not use the free running salts that do not cake as they are contaminated with bone ash or starch or some fine powder that does not dissolve in water and so the solution is milky. Use ordinary salt which is apt to cake but which is pure. Fill the lower third of the shade with this solution. Then allow water to trickle very slowly down the sides so as to float on the salt water below. Then fixing a funnel with a tap so that the mouth of the funnel is rather above the salt solution, allow a liquid called orthotoluidine to flow slowly out. This liquid is of a beautiful red colour and at a temperature of about 70° F. it has a density intermediate between that of water and the salt solution, and so a great spherical drop two or three inches in diameter may be formed and detached from the funnel and remain at rest in the vessel. If the temperature rises it goes up a little and the reverse if the temperature falls. Mr. Darling has also described an experiment in which a vessel of hot water kept hot from below at between 170° and 180° F. is used and into which aniline is poured At a temperature of about 145° F. aniline has the same density as water but as it expands more with rise of temperature it is lighter than water when hotter and heavier when cooler. The aniline on the surface becomes cooler and presently it gathers itself into a great pendent drop which breaks away from the surface and goes to the bottom. There it gets warmed and soon a great inverted drop forms and presently it breaks away and rises to the surface, and so the process goes on indefinitely. It is interesting to see the slow breaking off of the drops and the formation of small intermediate drops, about which I have more to say later.

Mr. Darling has also described a curious movement which occurs among the thin circular spots of these liquids which float upon clean water, a movement which is more marked if the liquids are not pure. While any spot of liquid may remain circular and quiescent for a time it is subject to convulsions which cause it to assume kidney shapes or to split into two or more. When there are a large number of spots the agitation is perpetual and is even more remarkable when the phenomenon is projected on a screen. Like the movements of camphor already described these are instantly destroyed by a trace of oil or soap applied to the surface of the water. The disks immediately then thicken into smaller lenses and remain at rest.

Petroleum, or paraffin as it is often called, will not mix with water but separates and floats on the top. If however some soap is dissolved in the water then the surface tension of the solution is so much less that if it separates at all it does so much more slowly.   Aphides and other insect pests do not like soft soap or petroleum, and so a mixture is useful for spraying trees.   If the petroleum separated from the liquid freely and got sprayed on to the tree by itself the tree might suffer as much as the insects, but as it remains diffused in the liquid the tree is not damaged.

The chief result that I have endeavoured to make clear so far is this.   The outside of a liquid acts as if it were an elastic skin, which will, as far as it is able, so mould the liquid within it that its surface shall be as small as possible.   Generally the weight of liquids, especially when there is a large quantity, is too much for the feebly elastic skin, and its power may not be noticed.   The disturbing effect of weight is got rid of by immersing one liquid in another which is equally heavy with which it does not mix, and it is hardly noticed when very small drops are examined, or when a bubble is blown, for in these cases the weight is almost nothing, while the elastic power of the skin is just as great as ever.   Different liquids have skins of different degrees of strength.   When liquids which do not mix are brought into contact with one another curious movements may result.

Soap-films, their Tension and Curvature

I have not as yet by any direct experiment shown that a soap-film or bubble is really elastic, like a piece of stretched india-rubber.

Before making any experiments, however, let us consider what sort of forces we are likely to have to investigate.   I have already stated that in the case of pure water the forces pulling in opposite directions across any line one inch long amount to the weight of $3\frac{1}{4}$ grains.

This is most easily determined by measuring the height that clean water rises in a narrow clean glass tube.   It is very commonly thought, as a soap solution will allow bubbles to be blown whereas clean water will not, that therefore the elastic strength of a soap solution must be greater.   It is really just the other way about and this may be seen at once by observing the rise of a soap

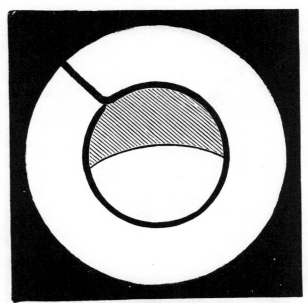

FIG. 20.

solution in the same tube that has been used for water. The soap solution rises to only about one-third of the height.   As such a solution need not be appreciably denser than water this shows that in the case of soap solution the pull is a little over one grain to the inch if in the case of water it is $3\frac{1}{4}$ grains to the inch.

Now a soap-bubble consists of a thin layer of liquid with two surfaces and each is tense or contractile to the extent of just over one grain to the inch; the bubble therefore is contractile to the extent of a little over two grains to the inch and that it is pulling at anything to which it is attached may easily be shown in many ways. Perhaps the easiest way is to tie a thread across a ring rather loosely, and then to dip the ring into soap water. On taking it out there is a film stretched over the ring, in which the thread moves about quite freely, as you can

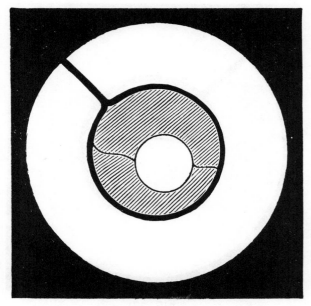

FIG. 21.

see upon the screen. But if I break the film on one side, then immediately the thread is pulled by the film on the other side as far as it can go, and it is now tight (Fig. 20). You will also notice that it is part of a perfect circle, because that form makes the space on one side as great, and therefore on the other side, where the film is, as small, as possible. Or again, in this second ring the thread is double for a short distance in the middle. If

I break the film between the threads they are at once
pulled apart, and are pulled into a perfect circle (Fig.
21), because that is the form which makes the space
within it as great as possible, and therefore leaves the
space outside it as small as possible. You will also
notice, that though the circle will not allow itself to be
pulled out of shape, yet it can move about in the ring

FIG. 22.

quite freely, because such a movement does not make
any difference to the size of the space outside it.

I have now blown a bubble upon a ring of wire. I
shall hang a small ring upon it, and to show more clearly
what is happening, I shall blow a little smoke into the
bubble. Now that I have broken the film inside the
lower ring, you will see the smoke being driven out and
the ring lifted up, both of which show the elastic nature
of the film. Or again, I have blown a bubble on the end

of a wide pipe; on holding the open end of the pipe to a candle flame, the outrushing air blows out the flame at once, which shows that the soap-bubble is acting like an elastic bag. Actually in this experiment the carbonic acid from the lungs assists greatly in putting out the candle, I have however blown out a freshly lighted candle with pure air in this way (Fig. 22). You now see that, owing to the

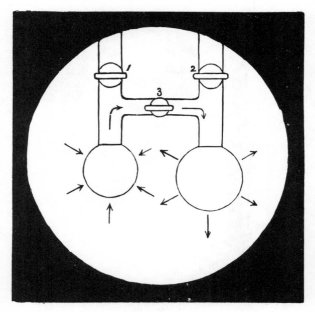

FIG. 23.

elastic skin of a soap-bubble, the air or gas inside is under pressure and will get out if it can. Which would you think would squeeze the air inside it most, a large or a small bubble? We will find out by trying, and then see if we can tell why. You now see two pipes each with a tap. These are joined together by a third pipe in which there is a third tap. I shall first blow one bubble and shut it off with the tap 1 (Fig. 23), and then the other, and shut it off with the tap 2. They are now nearly equal in size, but the air cannot yet pass from one to the other because the tap 3 is turned off. Now if the pressure in the larger one is greater it will blow air into the other when I open this tap, until they are equal in size ; if, on the other hand, the pressure in the smaller one

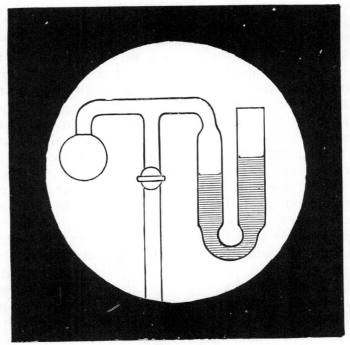

FIG. 24.

is greater, it will blow air into the larger one, and will itself get smaller until it has quite disappeared. We will now try the experiment. You see immediately that I open the tap 3 the small bubble shuts up and blows out the large one, thus showing that there is a greater pressure in a small than in a large bubble. The directions in which the air and the bubble move are indicated in the figure by arrows. I want you particularly to notice and remember this, because this is an experiment on which a great deal depends. To impress this upon your memory I shall show the same thing in another way. There is in front of the lantern a little tube shaped like a U half filled with water. One end of the U is joined to a pipe on which a bubble can be blown (Fig. 24). You will now be able to see how the pressure changes as the bubble increases in size, because the water will be displaced more when the pressure is more, and less when it is less. Now that there is a very small bubble, the pressure as measured by the water is about one quarter of an inch on the scale. The bubble is growing and the pressure indicated by the water in the gauge is falling, until, when the bubble is double its former size, the pressure is only half what it was ; and this is always true, the smaller the bubble the greater the pressure. As the film is always stretched with the same force, whatever size the bubble is, it is clear that the pressure inside can only depend upon the curvature of a bubble. In the case of lines, our ordinary language tells us, that the larger a circle is the less is its curvature ; a piece of a small circle is said to be a quick or a sharp curve, while a piece of a great circle is only slightly curved ; and if you take a piece of a very large circle indeed, then you cannot tell it from a straight line, and you say it is not curved at all. With a part of the surface of a ball it is just the same—the larger the ball the less it is curved : and if the ball is as large as the earth, *i. e.* 8000 miles across, you cannot tell a small piece of it from a true plane. Level water is part of such a surface, and you know that still water in a basin appears perfectly flat, though in a very large lake or the sea you can see that it is curved. We have seen that in large bubbles the pressure is little and the curvature is little, while in small bubbles the pressure is great and the curvature is great. The pressure and the curvature rise and fall together.

We have now learnt the lesson which the experiment of the two bubbles, one blown out by the other, teaches us.

A ball or sphere is not the only form which you can give to a soap-bubble.   If you take a bubble between two rings, you can pull it out until at last it has the shape of a round straight tube or cylinder as it is called.   We have spoken of the curvature of a ball or sphere ; now what is the curvature of a cylinder ?   Looked at sideways, the edge of the wooden cylinder upon the table appears straight, *i. e.* not curved at all ; but looked at from above it appears round, and is seen to have a definite curvature (Fig. 25).   What then is the curvature of the surface of a cylinder ?   We have seen that the pressure in a bubble

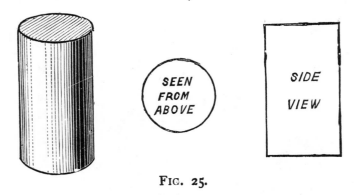

FIG. 25.

depends upon the curvature when they are spheres, and this is true whatever shape they have.   If, then, we find what sized sphere will produce the same pressure upon the air inside that a cylinder does, then we shall know that the curvature of the cylinder is the same as that of the sphere which balances it.   Now at each end of a short tube I shall blow an ordinary bubble, but I shall pull the lower bubble by means of another tube into the cylindrical form, and finally blow in more or less air until the sides of the cylinder are perfectly straight.   That is now done (Fig. 26), and the pressure in the two bubbles

must be exactly the same, as there is a free passage of air
between the two.   On measuring them you see that the
sphere is exactly double the cylinder in diameter.   But
this sphere has only half the curvature that a sphere
half its diameter would have.   Therefore the cylinder,
which we know has the same curvature that the large
sphere has, because the two balance, has only half the

Fig. 26.

curvature of a sphere of its own diameter, and the pressure in it is only half that in a sphere of its own
diameter.

I must now make one more step in explaining this
question of curvature.   Now that the cylinder and sphere
are balanced I shall blow in more air, making the
sphere larger; what will happen to the cylinder?   The
cylinder is, as you see, very short; will it become blown
out too, or what will happen?   Now that I am blowing

in air you see the sphere enlarging, thus relieving the pressure; the cylinder develops a waist, it is no longer a cylinder, the sides are curved inwards. As I go on blowing and enlarging the sphere, they go on falling inwards, but not indefinitely. If I were to blow the

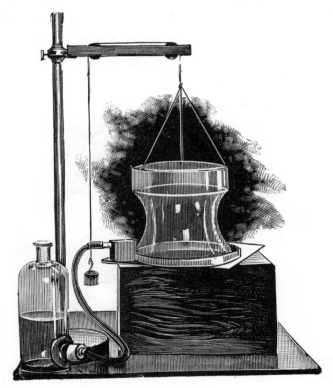

FIG. 27.

upper bubble till it was of an enormous size the pressure would become extremely small. Let us make the pressure nothing at all at once by simply breaking the upper bubble, thus allowing the air a free passage from the inside to the outside of what was the cylinder. Let me repeat this experiment on a large scale. I have two

large glass rings, between which I can draw out a film of
the same kind.   Not only is the outline of the soap-film
curved inwards, but it is exactly the same as the smaller
one in shape (Fig. 27).   As there is now no pressure
there ought to be no curvature, if what I have said is
correct.   But look at the soap-film.   Who would venture
to say that that is not curved? and yet we had satisfied
ourselves that the pressure and the curvature rose and fell

FIG. 28.

together.   We now seem to have come to an absurd
conclusion.   Because the pressure is reduced to nothing
we say the surface must have no curvature, and yet a glance
is sufficient to show that the film is so far curved as to
have a most elegant waist.   Now look at the plaster model
on the table, which is a model of a mathematical figure
which also has a waist.

Let us therefore examine this cast more in detail.   I
have a disc of card which has exactly the same diameter
as the waist of the cast.   I now hold this edgeways

against the waist (Fig. 28), and though you can see that it does not fit the whole curve, it fits the part close to the waist perfectly. This then shows that this part of the cast would appear curved inwards if you looked at it sideways, to the same extent that it would appear curved outwards if you could see it from above. So considering the waist only, it is curved both towards the inside and also away from the inside according to the way you look at it, and to the same extent. The curvature inwards would make the pressure inside less, and the curvature outwards would make it more, and as they are equal

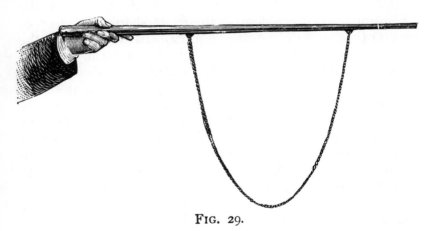

Fig. 29.

they just balance, and there is no pressure at all. If we could in the same way examine the bubble with the waist, we should find that this was true not only at the waist, but at every part of it. Any curved surface like this which at every point is equally curved opposite ways, is called a surface of no curvature, and so what seemed an absurdity is now explained. Now this surface, which is the only one of the kind symmetrical about an axis except a flat surface, is called a catenoid, because it is like a chain, as you will see directly, and, as you know, *catena* is the Latin for a chain. I shall now hang a chain in a loop from a level stick, and throw a strong light, upon it, so that you can see it well (Fig. 29). This is exactly the same shape as the side of a bubble drawn out between two rings, and open at the end to the air.

If in any bent surface the curvatures at any part as measured along any two lines at right angles to one another are not equal and opposite as they are in the catenoid we have just discussed, then if that surface is tense as is the surface of water there will have to be a greater pressure on the side that is most concave and this will be simply proportional to the difference of the two curvatures. This consideration affords a key to the solution of the problem of the exact shape of a drop of water (Fig. 2), or of alcohol (Fig. 12). In these cases as the interior of the surface is filled with liquid the pressure within the liquid gets steadily greater from above downwards, just as the pressure keeps on getting greater with greater depth in the sea. The form of the drop then is such that at any level the total curvature as defined above, that is the sum or the difference of the curvatures as measured in two directions at right angles to one another (sum if their centres are on the same side or difference if on opposite sides), is proportional to the distance below the free water or the alcohol level. Water is a heavier liquid which would alone make the drops smaller, on the other hand its skin is stronger than that of alcohol in a still higher degree and so its drops are larger. A comparison of Figs. 2 and 12 will show how great the difference is.

We have found that the pressure in a short cylinder gets less if it begins to develop a waist, and greater if it begins to bulge. Let us therefore try and balance one with a bulge against another with a waist. Immediately that I open the tap and let the air pass, the one with a bulge blows air round to the one with a waist and they both become straight. In Fig. 30 the direction of the movement of the air and of the sides of the bubble is indicated by arrows.   Let us next try the same experiment with a pair of rather longer cylinders, say about twice as long as they are wide.   They are now ready, one with a bulge and one with a waist.   Directly I open the tap, and let the air pass from one to the other, the one with a waist blows out the other still more (Fig. 31),

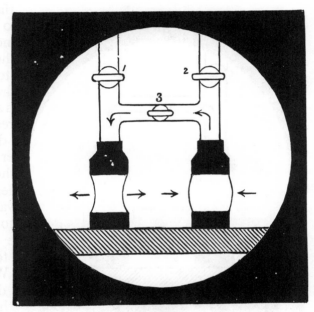

FIG. 30.

until at last it has shut itself up.   It therefore behaves exactly in the opposite way that the short cylinder did. If you try pairs of cylinders of different lengths you will find that the change occurs when they are just over one and a half times as long as they are wide.   Now if you imagine one of these tubes joined on to the end of the other, you will see that a cylinder more than about three times as long as it is wide cannot last more than a moment ; because if one end were to contract ever so little the pressure there would increase, and the narrow end would blow air into the wider end (Fig. 32), until the sides of the narrow end met one another. The exact length of the longest cylinder that is stable, is a little more than three diameters. The cylinder just becomes unstable when its length is equal to its circumference, and this is $3\frac{1}{7}$ diameters almost exactly.

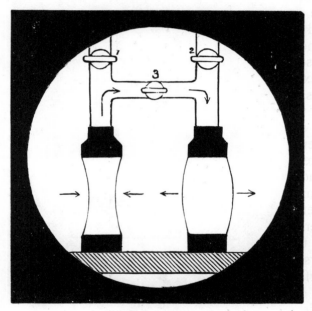

FIG. 31.

I shall gradually separate these rings, keeping up a supply of air, and you will see that when the tube gets nearly three times as long as it is wide it is getting very difficult to manage, and then suddenly it grows a waist nearer one end than the other, and breaks off forming a pair of separate and unequal bubbles.

As a soap-bubble is tense and always moves so as to become as small in surface as the conditions determined

by the air that it contains or by its attachment to solid
supports will allow, it will be pretty obvious that it
enables us to find out when a change of form increases
or decreases the total surface. For instance, in the case
of the cylinder just considered, supported between two
rings and containing so much air, if the length of the
cylinder is less than $3\frac{1}{7}$ diameters then a narrowing at
one end and the necessary widening
at the other to accommodate the displaced air makes the total surface
become greater, and this we know
because the soap-bubble of that form
can exist. One longer than $3\frac{1}{7}$ diameters cannot exist, and therefore we
know that any movement tending to
form a waist and to bulge at the
two ends however small makes the
surface smaller, and so the soap-bubble
will not go back, but by continuing
the movement makes the surface
smaller still and so on, until it has
broken through as we have seen. Just
close to the critical length of $3\frac{1}{7}$ diameters there is excessively little change
of surface for a small movement of the
kind considered, and so the soap-
bubble either resists or encourages
such movement with very feeble influence, and such bubbles are said to
be very slightly stable or only just

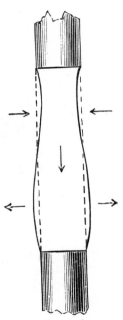

FIG. 32.

unstable as the case may be, and the very slightly
stable bubble may be used to detect forces acting
on the gas within it which would not be noticed in a
bubble of more stable form, such for instance as a
spherical bubble. I have blown a spherical bubble
with oxygen gas and placed it between the poles
of an electro-magnet, that is iron which only becomes

magnetic when I allow an electric current to pass in the wire round it (Fig. 33). The bubble and magnet can be seen on the screen, and you can hear the tap of the key by which the electric current is allowed to pass. There must be some movement of the bubble because oxygen gas is slightly magnetic, but I doubt if any one can detect the movement. Now by means of a stand with moveable rings I can blow another bubble of the same

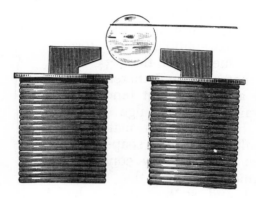

Fig. 33.

gas and draw it out into a cylindrical bubble of very nearly the critical length (Fig. 34). The instant that you hear the click of the key the magnetic pull on the oxygen gas enables it to overcome the feeble resistance of the nearly unstable bubble, and in a moment, too quick for the eye to follow it, the bubble has separated into two (Fig. 35).

Liquid Cylinders and Jets

If now you have a cylinder of liquid of great length suddenly formed and left to itself, it clearly cannot retain that form. It must break up into a series of drops. Unfortunately the changes go on so quickly in a falling

stream of water that no one by merely looking at it
could follow the movements of the separate drops, but I
hope to be able to show to you in two or three ways
exactly what is happening. You may remember that we

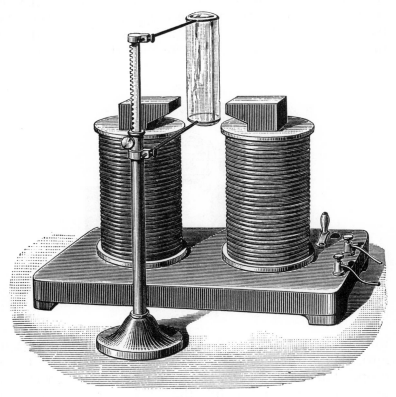

FIG. 34.

were able to make a large drop of one liquid in another,
because in this way the effect of the weight was neutralized, and as large drops oscillate or change their shape
much more slowly than small, it is more easy to see what
is happening. I have in this glass box water coloured
blue on which is floating petroleum, made heavier by

mixing with it a bad-smelling and dangerous liquid called
bisulphide of carbon.

The water is only a very little heavier than the mixture.
If I now dip a pipe into the water and let it fill, I can
then raise it and allow drops to form slowly.   Drops as
large as a shilling are now forming, and when each one
has reached its full size, a neck forms above it, which is
drawn out by the falling drop into a little cylinder.   You

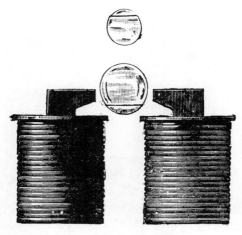

FIG. 35.

will notice that the liquid of the neck has gathered itself
into a little drop which falls away just after the large
drop.   The action is now going on so slowly that you can
follow it.   If I again fill the pipe with water, and
this time draw it rapidly out of the liquid, I shall leave
behind a cylinder which will break up into balls as you
can easily see (Fig. 36).   I should like now to show you,
as I have this apparatus in its place, that you can blow
bubbles of water containing petroleum in the petroleum
mixture, and you will see some which have other bubbles
and drops of one or other liquid inside again.   One of
these compound bubble drops is now resting stationary

Fig. 36.

on a heavier layer of liquid, so that you can see it all the better (Fig. 37). If I rapidly draw the pipe out of the box I shall leave a long cylindrical bubble of water containing petroleum, and this, as was the case with the water-cylinder, slowly breaks up into spherical bubbles. For all these experiments the liquids orthotoluidine and water would be more convenient.

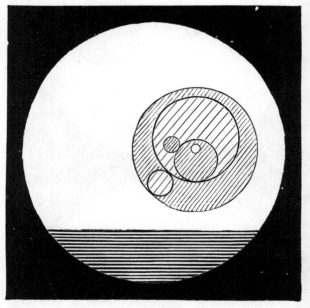

FIG. 37.

One of the most beautiful bubbles of one liquid in another which can be produced is occasionally formed by accident. If a basin of water containing a few pounds of mercury is placed under a violently running water-tap the water and air carried down into the mercury cause mercury bubbles to form and float to the surface. I have been able to float these into a second basin, where sometimes for a few seconds they look like shining balls of pure silver, perfect in form and polish. When they break a tiny globule of mercury alone remains, far more however than the liquid of a soap-bubble of the same size. I have obtained mercury bubbles up to about

FIG. 38.

¾ inch in diameter. M. Melsens, who first described these in 1845, found the upper part to be so thin as to be transparent and of a slatey-blue colour, a phenomenon which I have not noticed. This experiment is apt to be disastrous if it is done in a lead sink or in a sink with a lead drain-pipe, besides being wasteful of mercury. Care should be taken that there is always a larger vessel in which the basin stands to catch the overflowing mercury.

Having shown that a very large liquid cylinder breaks up regularly into drops, I shall next go to the other extreme, and take as an example an excessively fine cylinder. You see a photograph of a diadema spider on her geometrical web (Fig. 38). If I had time I should like to tell you how the spider goes to work to make this beautiful structure, and a great deal about these wonderful creatures, but I must do no more than show you that there are two kinds of web—those that point outwards, which are hard and smooth, and those that go round and round, which are very elastic, and which are covered with beads of a sticky liquid. Now there are in a good web over a quarter of a million of these beads which catch the flies for the spider's dinner. A spider makes a whole web in an hour, and generally has to make a new one every day. She would not be able to go round and stick all these in place, even if she knew how, because she would not have time. Instead of this she makes use of the way that a liquid cylinder breaks up into beads as follows. She spins a thread, and at the same time wets it with a sticky liquid, which of course is at first a cylinder. This cannot remain a cylinder, but breaks up into beads, as the photograph taken with a microscope from a real web beautifully shows (Fig. 39). You see the alternate large and small drops, and sometimes you even see extra small drops between these again. In order that you may see exactly how large these beads really are, I have placed alongside a scale of thousandths of an inch, which I photographed at the same time. To prove to you that this is what happens, I shall now show you a web that I have made myself by stroking a quartz fibre with a straw dipped in castor-oil. The same alternate large and small beads are again visible just as perfect as they were in the spider's web. In fact it is impossible to distinguish between one of my beaded webs and a spider's by looking at them. You might say that a large cylinder of water in oil, or a microscopic cylinder on a thread, is not the same as an ordinary jet of water, and that you would like to see if it behaves as I have described. The next photograph (Fig. 40), taken by the light of an instantaneous electric spark, and magnified three and a quarter times, shows a fine column of water falling from a jet. You will now see that it is at first a cylinder, that as it goes down necks and bulges begin to form, and at last beads separate, and you can see the little drops as well. The beads also vibrate, becoming alternately long and wide, and there can be no doubt that the sparkling portion of a jet, though it appears continuous, is really made up of beads which pass so rapidly before the eye that it is impossible to follow them. (I should explain that for a reason which will appear later, I made a loud note by whistling into a key at the time that this photograph was taken.)

Lord Rayleigh has shown that in a stream of water one twenty-fifth of an inch in diameter, necks impressed upon the stream, even though imperceptible,

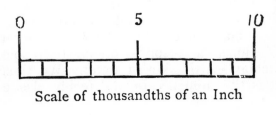

0   5   10

Scale of thousandths of an Inch

Fig. 39.

develop a thousandfold in depth every fortieth of a second, and thus it is not difficult to understand that in such a stream the water is already broken through before it has fallen many inches. He has also shown that free water-drops vibrate at a rate which may be found as follows. A drop two inches in diameter makes one complete vibration in one second. If the diameter is reduced to one quarter of its amount, the time of vibration will be reduced to one-eighth, or if the diameter is reduced to one-hundredth, the time will be reduced to one-thousandth, and so on. The same relation between the diameter and the time of breaking up applies also to cylinders. We can at once see how fast a bead of water the size of one of those in the spider's web would vibrate if pulled out of shape, and let go suddenly. If we take the diameter as being one eight-hundredth of an inch, and it is really even smaller, then the bead would have a diameter of one-sixteen-hundredth of a two-inch bead, which makes one vibration in one second. It will therefore vibrate sixty-four thousand times as fast, or sixty-four thousand times a second. Water-drops the size of the little beads, with a diameter of rather less than one three-thousandth of an inch, would vibrate half a million times a second, under the sole influence of the feebly elastic skin of water. We thus see how powerful is the influence of the feebly elastic

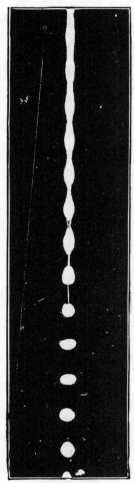

FIG. 40.

water-skin on drops of water that are sufficiently small.

I shall now cause a small fountain to play, and shall allow the water as it falls to patter upon a sheet of paper. You can see both the fountain itself and its shadow upon the screen. You will notice that the water comes out of the nozzle as a smooth cylinder, that it presently begins to

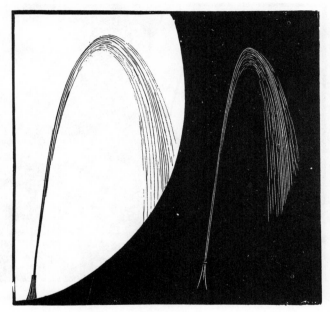

FIG. 41.

glitter, and that the separate drops scatter over a great space (Fig. 41). Now why should the drops scatter? All the water comes out of the jet at the same rate and starts in the same direction, and yet after a short way the separate drops by no means follow the same paths. Now instead of explaining this, and then showing experiments to test the truth of the explanation, I shall reverse the usual order, and show one or two experiments first, which I think you will agree are almost like magic.

You now see the water of the jet scattering in all directions, and you hear it making a pattering sound on the paper on which it falls.   I take out of my pocket a stick of sealing-wax and instantly all is changed, even though I am some way off and can touch nothing.   The water ceases to scatter ; it travels in one continuous line

Fig. 42.

(Fig. 42), and falls upon the paper making a loud rattling noise which must remind you of the rain of a thunderstorm.   I come a little nearer to the fountain and the water scatters again, but this time in quite a different way.   The falling drops are much larger than they were before.   Directly I hide the sealing-wax the jet of water recovers its old appearance, and as soon as the sealing-wax is taken out it travels in a single line again.

Now instead of the sealing-wax I shall take a smoky flame easily made by dipping some cotton-wool on the end of a stick into benzene, and lighting it. As long as the flame is held away from the fountain it produces no effect, but the instant that I bring it near so that the water passes through the flame, the fountain ceases to scatter; it all runs in one line and falls in a dirty black stream upon the paper. Ever so little oil fed into the

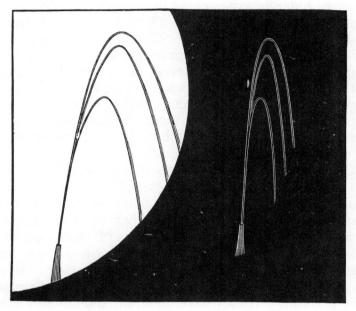

FIG. 43.

jet from a tube as fine as a hair does exactly the same thing.

I shall now set a tuning-fork sounding at the other side of the table. The fountain has not altered in appearance. I now touch the stand of the tuning-fork with a long stick which rests against the nozzle. Again the water gathers itself together even more perfectly than before, and the paper upon which it falls is humming out a note which is the same as that produced by the tuning-fork. If I alter the rate at which the water flows you

will see that the appearance is changed again, but it is
never like a jet which is not acted upon by a musical
sound. Sometimes the fountain breaks up into two or
three and sometimes many more distinct lines, as though
it came out of as many tubes of different sizes and
pointing in slightly different directions (Fig. 43). The
effect of different notes could be very easily shown if any
one were to sing to the piece of wood by which the jet
is held. I can make noises of different pitches, which
for this purpose are perhaps better than musical notes,
and you can see that with every new noise the fountain
puts on a different appearance. You may well wonder
how these trifling influences—sealing-wax, the smoky
flame, oil or the more or less musical noise—should produce this mysterious result, but the explanation is not
so difficult as you might expect.

Consider what I have said about a liquid cylinder. If it
is a little more than three times as long as it is wide, it
cannot retain its form; if it is made very much more
than three times as long, it will break up into a series of
beads. Now, if in any way a series of necks could be
developed upon a cylinder which were less than three
diameters apart, some of them would tend to heal up,
because a piece of a cylinder less than three diameters
long is stable. If they were about three diameters apart,
the form being then unstable, the necks would get more
pronounced in time, and would at last break through, so
that beads would be formed. If necks were made at
distances more than three diameters apart, then the
cylinder would go on breaking up by the narrowing of
these necks, and it would most easily break up into drops
when the necks were just four and a half diameters
apart. In other words, if a fountain were to issue from a
nozzle held perfectly still, the water would most easily
break into beads at the distance of four and a half
diameters apart, but it would break up into a greater
number closer together, or a smaller number further

apart, if by slight disturbances of the jet very slight waists were impressed upon the issuing cylinder of water. When you make a fountain play from a jet which you hold as still as possible, there are still accidental tremors of all kinds, which impress upon the issuing cylinder slightly narrow and wide places at irregular distances, and so the cylinder breaks up irregularly into drops of different sizes and at different distances apart. Now these drops, as they are in the act of separating from one another, and are drawing out the waist, as you have seen, are being pulled for the moment towards one another by the elasticity of the skin of the waist ; and, as they are free in the air to move as they will, this will cause the hinder one to hurry on, and the more forward one to lag behind, so that unless they are all exactly alike both in size and distance apart they will many of them bounce together before long. You would expect when they hit one another afterwards that they would join, but I shall be able to show you in a moment that they do not ; they act like two india-rubber balls, and bounce away again. Now it is not difficult to see that if you have a series of drops of different sizes and at irregular distances bouncing against one another frequently, they will tend to separate and to fall, as we have seen, on all parts of the paper down below. What did the sealing-wax, the smoky flame or the oil do? and what can the musical sound do to stop this from happening? Let me first take the sealing-wax. A piece of sealing-wax rubbed on your coat is electrified, and will attract light bits of paper up to it. The sealing-wax acts electrically on the different water-drops, causing them to attract one another, feebly, it is true, but with sufficient power where they meet to make them break through the air-film between them and join. To show that this is not imaginary, I have now in front of the lantern two fountains of clean water coming from separate bottles, and you can see that they bounce apart perfectly (Fig. 44). To show

that they do really bounce, I have coloured the water in
the two bottles differently.   The sealing-wax is now in my
pocket ; I shall retire to the other side of the room, and
the instant it appears the jets of water coalesce (Fig. 45).
This may be repeated as often as you like, and it never
fails.   These two bouncing jets provide one of the
most delicate tests for the presence of electricity that

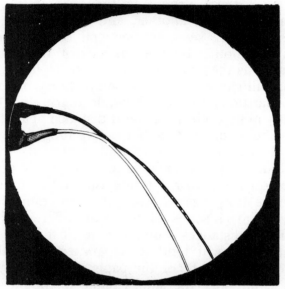

FIG. 44.

exist.   You are now able to understand the first experiment.   The separate drops which bounced away from one
another, and scattered in all directions, are unable to
bounce when the sealing-wax is held up, because of its
electrical action.   They therefore unite, and the result is,
that instead of a great number of little drops falling all
over the paper, the stream pours in a single line, and
great drops, such as you see in a thunder-storm, fall on the
top of one another.   There can be no doubt that it is

for this reason that the drops of rain in a thunder-storm are so large. This experiment and its explanation are due to Lord Rayleigh.

The bouncing jets are so sensitive that they may be used to receive wireless messages if the liquid used is dilute sulphuric acid so as to conduct electricity, and the receiving circuit is broken by the air space between the

FIG. 45.

jets. I have set up a little recorder, the tail of which is hit by the coalesced jet so as to move its head into contact with one of the individual streams which immediately causes their separation, and the jets are ready for another signal. It is interesting to notice when this is set up in a large hall at the other end of which the transmitting sparks are made that the movement of the recorder, as seen upon the screen, anticipates the sound of the spark; the fact being that the action on the jets is at the same moment as the spark, but the sound comes along perhaps a tenth of a second later.

The smoky flame, as shown by Mr. Bidwell, also makes the jet of water flow in a single line. The reason probably is that the dirt breaks through the air-film, just as dust in the air will make the two fountains join as they did when they were electrified. However, it is possible that oily matter condensed on the water may have something to do with the effect observed, because oil alone acts quite as well as a flame, but the action of oil in this case, as when it smooths a stormy sea, is not by any means so easily understood.

When I held the sealing-wax closer, the drops coalesced in the same way; but they were then so much more electrified that they repelled one another as similarly electrified bodies are known to do, and so the electrical scattering was produced.

You possibly already see why the tuning-fork made the drops follow in one line, but I shall explain. A musical note is, as is well known, caused by a rapid vibration; the more rapid the vibration the higher is the pitch of the note. For instance, I have a tooth-wheel which I can turn round very rapidly if I wish. Now that it is turning slowly you can hear the separate teeth knocking against a card that I am holding in the other hand. I am now turning faster, and the card is giving out a note of a low pitch. As I make the wheel turn faster and faster, the pitch of the note gradually rises, and it would, if I could only turn fast enough, give so high a note that we should not be able to hear it. A tuning-fork vibrates at a certain definite rate, and therefore gives a definite note. The fork now sounding vibrates 128 times in every second. The nozzle, therefore, is made to vibrate, but almost imperceptibly, 128 times a second, and to impress upon the issuing cylinder of water 128 imperceptible waists every second. Now it just depends what size the jet is, and how fast the water is issuing, whether these waists are about four and a half diameters apart in the cylinder. If the jet is larger, the water must pass more quickly, or under a greater pressure, for this to be the case ; if the jet is finer, a smaller speed will be sufficient. If it should happen that the waists so made are anywhere about four diameters apart, then even though they are so slightly developed that if you had an exact drawing of them, you would not be able to detect the slightest change of diameter, they will grow at a great speed, and therefore the water column will break up regularly, every drop will be like the one behind it, and like the one in front of it, and not all different, as is the case when the breaking of the water merely depends upon accidental tremors. If the drops then are all alike in every respect, of course they all follow the same path, and so appear to fall in a continuous stream. If the waists are about four and a half diameters apart, then the jet will break up most easily ; but it will, as I have said, break up under the influence of a considerable range of notes, which cause the waists to be formed at other distances, provided they are more than three diameters apart. If two notes are sounded at the same time, then very often each will produce its own effect, and the result is the alternate formation of drops of different sizes, which then make the jet divide into two separate streams. In this way, three, four, or even many more distinct streams may be produced.

I can now show you photographs of some of these musical fountains, taken by the instantaneous flash of an electric spark, and you can see the separate paths described by the drops of different sizes (Fig. 46). In one photograph there are eight distinct fountains all breaking from the same jet, but following quite distinct paths, each of which is clearly marked out by a perfectly regular series of drops. You can also in these photographs see drops actually in the act of bouncing against

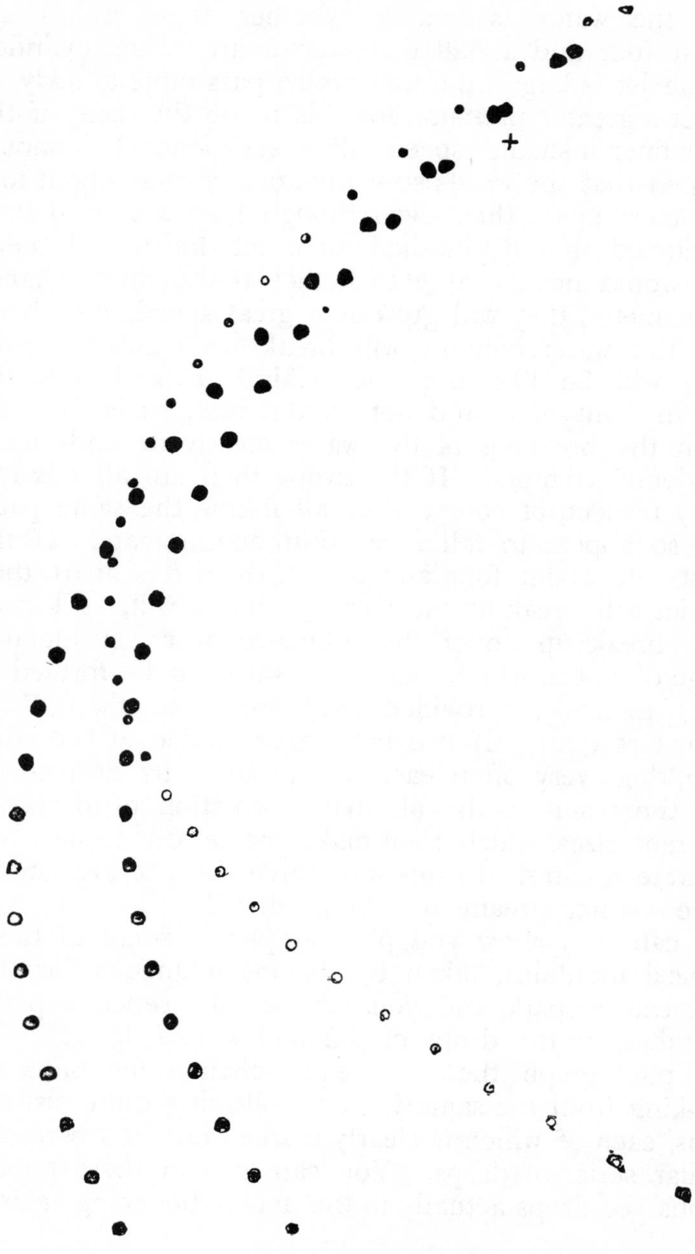

FIG. 46.

one another, and flattened when they meet, as if they were india-rubber balls. In the photograph now upon the screen the effect of this rebound, which occurs at the place marked with a cross, is to hurry on the upper and more forward drop, and to retard the other one, and so to make them travel with slightly different velocities and directions. It is for this reason that they afterwards follow distinct paths. The smaller drops had no doubt been acted on in a similar way, but the part of the fountain where this happened was just outside the photographic plate, and so there is no record of what occurred. The very little drops of which I have so often spoken are generally thrown out from the side of a fountain of water under the influence of a musical sound, after which they describe regular little curves of their own, quite distinct from the main stream. They, of course, can only get out sideways after one or two bouncings from the regular drops in front and behind. You can easily show that they are really formed below the place where they first appear, by taking a piece of electrified sealing-wax and holding it near the stream close to the nozzle and gradually raising it. When it comes opposite to the place where the little drops are really formed, it will act on them more powerfully than on the large drops, and immediately pull them out from a place where the moment before none seemed to exist. They will then circulate in perfect little orbits round the sealing-wax, just as the planets do round the sun ; but in this case, being met by the resistance of the air, the orbits are spirals, and the little drops after a few revolutions ultimately fall upon the wax, just as the planets would fall into the sun after many revolutions, if their motion through space were interfered with by friction of any kind.

There is only one thing needed to make the demonstration of the behaviour of a musical jet complete, and that is, that you should yourselves see these drops in their different positions in an actual fountain of water.

Now if I were to produce a powerful electric spark, then it is true that some of you might for an instant catch sight of the drops, but I do not think that most would see anything at all. But if, instead of making merely one flash, I were to make another when each drop had just travelled to the position which the one in front of it occupied before, and then another when each drop had moved on one place again, and so on, then all the drops, at the moments that the flashes of light fell upon them, would occupy the same positions, and thus all these drops would appear fixed in the air, though of course they really are travelling fast enough. If, however, I do not quite succeed in keeping exact time with my flashes of light, then a curious appearance will be produced. Suppose, for instance, that the flashes of light follow one another rather too quickly, then each drop will not have had quite time enough to get to its proper place at each flash, and thus at the second flash all the drops will be seen in positions which are just behind those which they occupied at the first flash, and in the same way at the third flash they will be seen still further behind their former places, and so on, and therefore they will appear to be moving slowly backwards; whereas if my flashes do not follow quite quickly enough, then the drops will, every time that there is a flash, have travelled just a little too far, and so they will all appear to be moving slowly forwards. Now let us try the experiment. There is the electric lantern sending a powerful beam of light on to the screen. This I bring to a focus with a lens, and then let it pass through a small hole in a piece of card. The light then spreads out and falls upon the screen. The fountain of water is between the card and the screen, and so a shadow is cast which is conspicuous enough. Now I place just behind the card a little electric motor, which will make a disc of card which has six holes near the edge spin round very fast. The holes come one after the other opposite the hole in the fixed card, and so at every turn six flashes of light are produced. When the card is turning $21\frac{1}{3}$ times a second, then the flashes will follow one another at the right rate. I have now started the motor, and after a moment or two I shall have obtained the right speed, and this I know by blowing through the holes, when a musical note will be produced, higher than the fork if the speed is too high, and lower than the fork if the speed is too low, and exactly the same as the fork if it is right.

To make it still more evident when the speed is exactly right, I have placed the tuning-fork also between the light and the screen, so that you may see it illuminated, and its shadow upon the screen. I have not yet allowed the water to flow, but I want you to look at the fork. For a moment I have stopped the motor, so that the light may be steady, and you can see that the fork is in motion because its legs appear blurred at the ends, where of course the motion is most rapid. Now the motor is started, and almost at once the fork appears quite different. It now looks like a piece of india-rubber, slowly opening and shutting, and now it appears quite still, but the noise it is making shows that it is not still by any means. The legs of the fork are vibrating, but the light only falls upon them at regular intervals, which correspond with their movement, and so, as I explained in the case of the water-drops, the fork appears perfectly still. Now the speed is slightly altered, and, as I have explained, each new flash of light, coming just too soon or just too late, shows the fork in a position which is just before or just behind that made visible by the previous flash. You thus see the fork slowly going through its evolutions, though of course in reality the legs are moving backwards and forwards 128 times a second. By looking at the fork or its shadow, you will therefore be able to tell whether the light is keeping exact time with the vibrations, and therefore with the water-drops.

Now the water is running, and you see all the separate

drops apparently stationary, strung like pearls or beads of silver upon an invisible wire (*see* Frontispiece). If I make the card turn ever so little more slowly, then all the drops will appear to march onwards slowly, and what is so beautiful—but I am afraid few will see this—each little drop may be seen to break off gradually, pulling out a waist which becomes a little drop, and then when the main drop is free it slowly oscillates, becoming wide and long, or turning over and over, as it goes on its way. If it so happens that a double or multiple jet is being produced, then you can see the little drops moving up to one another, squeezing each other where they meet and bouncing away again. Now the card is turning a little too fast and the drops appear to be moving backwards, so that it seems as if the water is coming up out of the tank on the floor, quietly going over my head, down into the nozzle, and so back to the water-supply of the place. Of course this is not happening at all, as you know very well, and as you will see if I simply try and place my finger between two of these drops. The splashing of the water in all directions shows that it is not moving quite so quietly as it appears. There is one more thing that I would mention about this experiment. Every time that the flashing light gains or loses one complete flash, upon the motion of the tuning-fork, it will appear to make one complete oscillation, and the water-drops will appear to move back or on one place.

A Water Telephone

I must now come to one of the most beautiful applications of these musical jets to practical purposes which it is possible to imagine, and what I shall now show are a few out of a great number of the experiments of Mr. Chichester Bell, cousin of Mr. Graham Bell, the inventor of the telephone.

To begin with I have a very small jet of water forced

through the nozzle at a great pressure, as you can see
if I point it towards the ceiling, as the water rises eight
or ten feet.   If I allow this stream of water to fall upon
an india-rubber sheet, stretched over the end of a tube
as big as my little finger, then the little sheet will be
depressed by the water, and the more so if the stream
is strong.   Now if I hold the jet close to the sheet the
smooth column of liquid will press the sheet steadily,
and it will remain quiet ; but if I gradually take the jet
further away from the sheet, then any waists that may
have been formed in the liquid column, which grow as
they travel, will make their existence perfectly evident.
When a wide part of the column strikes the sheet it will
be depressed rather more than usual, and when a narrow
part follows, the depression will be less.   In other words,
any very slight vibration imparted to the jet will be
magnified by the growth of waists, and the sheet of
india-rubber will reproduce the vibration, but on a magnified scale.   Now if you remember that sound consists
of vibrations, then you will understand that a jet is a
machine for magnifying sound.   To show that this is
the case I am now directing the jet on to the sheet, and
you can hear nothing ; but I shall hold a piece of wood
against the nozzle, and now, if on the whole the jet
tends to break up at any one rate rather than at any
other, or if the wood or the sheet of rubber will vibrate
at any rate most easily, then the first few vibrations
which correspond to this rate will be imparted to the
wood, which will impress them upon the nozzle and so
upon the cylinder of liquid, where they will become
magnified ; the result is that the jet immediately begins
to sing of its own accord, giving out a loud note (Fig. 47).

I will now remove the piece of wood.   On placing
against the nozzle an ordinary lever watch, the jolt
which is imparted to the case at every tick, though it
is so small that you cannot detect it, jolts the nozzle
also, and thus causes a neck to form in the jet of water

which will grow as it travels, and so produce a loud tick, audible in every part of this large room (Fig. 48). Now I want you to notice how the vibration is magnified by the action I have described. I now hold the nozzle close to the rubber sheet, and you can hear nothing. As I gradually raise it a faint echo is produced, which

FIG. 47.

gradually gets louder and louder, until at last it is more like a hammer striking an anvil than the tick of a watch.

I shall now change this watch for another which, thanks to a friend, I am able to use. This watch is a repeater, that is, if you press upon a knob it will strike, first the hour, then the quarters, and then the minutes. I think the water-jet will enable you all to hear what time it is. Listen! one, two, three, four; . . . . ting-tang, ting-tang; . . . one, two, three, four, five, six. Six minutes after half-past four. You notice that not only did you hear the number of strokes, but the jet faithfully reproduced the musical notes, so that you could distinguish one note from the others.

FIG. 48.

I can in the same way make the jet play a tune by simply making the nozzle rest against a long stick, which is pressed upon a musical-box. The musical-box is carefully shut up in a double box of thick felt, and you can hardly hear anything; but the moment that the nozzle is made to rest against the stick and the water is directed upon the india-rubber sheet, the sound of the box is loudly heard, I hope, in every part of the room. It is usual to describe a fountain as playing, but it is now evident that a fountain can even play a tune. There is, however, one peculiarity which is perfectly evident. The jet breaks up at certain rates more easily than at others, or, in other words, it will respond to certain sounds in preference to others. You can hear that as the musical-box plays, certain notes are emphasized in a curious way, producing much the same effect that follows if you lay a penny upon the upper strings of a horizontal piano.

Soap-films on Frames

Now, on returning to our soap-bubbles, you may remember that I stated that the catenoid and the plane were the only figures of revolution which had no curvature, and which therefore produced no pressure. There are plenty of other surfaces which are apparently curved in all directions and yet have no curvature, and which therefore produce no pressure; but these are not figures of revolution, that is, they cannot be obtained by simply spinning a curved line about an axis. These may be produced in any quantity by making wire frames of various shapes and dipping them in soap and water. On taking them out a wonderful variety of surfaces of no curvature will be seen. One such surface is that known as the screw-surface. To produce this it is only necessary to take a piece of wire wound a few times in an open helix (commonly called spiral), and to bend the two ends so as to meet a second wire passing down the centre. The screw-surface developed by dipping this frame in soap-water is well worth seeing (Fig. 49). It is impossible to give any idea of the perfection of the form in a figure, but fortunately this is an experiment which any one can easily perform.

It may be worth while to mention a curious relation-

ship between the screw surface and the catenoid of
revolution (Figs. 49 and 28) both of which are surfaces
of no curvature, which therefore may be realized with
a soap-film. You know that a flat piece of paper will
bend but will not stretch, and thus a sheet of paper may
be bent into the form of a cylinder or of a cone without
stretching any part of it. It cannot be bent into a
sphere or part of a sphere, as this would require that the

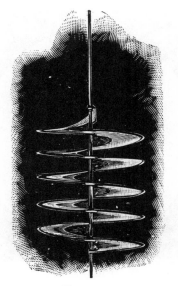

FIG. 49.

middle parts should be stretched or the outer parts com-
pressed, and this paper resists. Now by building up
on a turned and greased model of the catenoid a number
of strips of thin paper with paste so as to overlap and
cross one another and so make a catenoid of paper, the
curious relationship may be shown. When the paste
has dried cut the paper through with a knife along one
radial plane so as to get it off, and then, holding it by
the two cut ends of the waist, pull these apart, giving

them a twist at the same time, and then, when the waist is pulled out straight, the rest of the paper has bent itself without stretching at any part into a perfect screw surface, a double-bladed screw surface.

If a wire frame is made in the shape of the edges of any of the regular geometrical solids, very beautiful figures will be found upon them after they have been dipped in soap-water. In the case of the triangular

FIG. 50.

prism these surfaces are all flat, and at the edges where these planes meet one another there are always three meeting each other at equal angles (Fig. 50). This, owing to the fact that the frame is three-sided, is not surprising. After looking at this three-sided frame with three films meeting down the central line, you might expect that with a four-sided or square frame there would be four films meeting each other in a line down the middle. But it is a curious thing that it does not matter how irregular the frame may be, or how complicated a mass of froth may be, there can never be more than three films meeting in an edge, or more than four edges, or six films, meeting in a point. Moreover the films and edges can only meet one another at equal angles. If for a moment by any accident four films do meet in the same edge, or if the angles are not exactly equal, then the form, whatever it may be, is unstable; it cannot last, but the films slide over one another and never rest until they have settled down into a position in which the con-

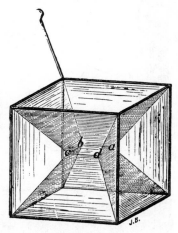

FIG. 51.

ditions of stability are fulfilled. As a consequence a cubical frame gives rise to the pattern shown in Fig. 51, in which the central square may be parallel to any one of the six faces of the cube, and the twelve other films so adjust themselves as to comply with the rule that all the angles must be angles of 120°. The general rule may be illustrated by a very simple experiment which you can easily try at home, and which you can now see projected upon the screen. There are two pieces of window-glass about half an inch apart, which form the sides of a sort of box into which some soap and water have been poured. On blowing through a pipe which is immersed in the water, a great number of bubbles are formed between the plates. If the bubbles are all large enough to reach across from one plate to the other, you will at once see that there are nowhere more than three films meeting one another, and where they meet the angles are all equal. The curvature of the bubbles makes it difficult to see at first that the angles really are all alike, but if you only look at a very short piece close to where they meet, and so

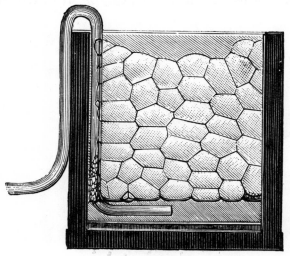

FIG. 52.

avoid being bewildered by the curvature, you will see that what I have said is true. You will also see, if you are quick, that when the bubbles are blown, sometimes four for a moment do meet, but that then the films at once slide over one another and settle down into their only possible position of rest (Fig. 52).

Soap -bubbles and Ether

The air inside a bubble is generally under pressure, which is produced by its elasticity and curvature. If the bubble would let the air pass through it from one side to the other of course it would soon shut up, as it did when a ring was hung upon one, and the film within the ring was broken. But there are no holes in a bubble, and so you would expect that a gas like air could not pass through to the other side. Nevertheless it is a fact that gases can slowly get through to the other side, and in the case of certain vapours the process is far more rapid than any one would think possible.

FIG. 53.

Ether produces a vapour which is very heavy, and which also burns very easily. This vapour can get to the other side of a bubble almost at once. It does so however not exactly by passing through it but by condensing on one side and evaporating on the other. I shall pour a little ether upon blotting-paper in this bell jar, and fill the jar with its heavy vapour. You can see that the jar is filled with something, not by looking at it, for it appears empty, but by looking at its shadow on the screen. Now I tilt it gently to one side, and you see something pouring out of it, which is the vapour of ether.

It is easy to show that this is heavy ; it is only necessary to drop into the jar a bubble, and so soon as the bubble meets the heavy vapour it stops falling and remains floating upon the surface as a cork does upon water (Fig. 53). Now let me test the bubble and see whether any of the vapour has passed to the inside.   I pick it up

FIG. 54.

out of the jar with a wire ring and carry it to a light, and at once there is a burst of flame. But this is not sufficient to show that the ether vapour has passed to the inside, because it might have condensed in sufficient quantity upon the bubble to make it inflammable. You remember that when I poured some of this vapour upon water (see p. 34), sufficient condensed to so weaken the water-skin that the frame of wire could get through to the other side.   However, I can see whether this is the true explanation or not by blowing a bubble on a wide pipe, and holding it in the vapour for a moment.   Now on removing it you notice that the bubble hangs like a heavy drop ; it has lost the perfect round-ness that it had at first, and this looks as if the vapour had found its way in, but this is made certain by bringing a light to the mouth of the tube, when the vapour, forced out by the elasticity of the bubble, catches fire and burns with a flame five or six inches long (Fig. 54).   You might also have noticed that when the bubble was re-

FIG. 55.

moved, the vapour inside it began to pass out again and fell away in a heavy stream, but this you could only see by looking at the shadow upon the screen.

Experiments with Soap-bubbles

You may have noticed when I made the drops of oil in the mixture of alcohol and water, that when they were brought together they did not at once unite ; they pressed against one another and pushed each other away if allowed, just as the water-drops did in the foun-tain of which I showed you a photograph.   You also may have noticed that the drops of water in the paraffin mixture bounced against one another, or if filled with the paraffin, formed bubbles in which often other small drops, both of water and paraffin, remained floating.

In all these cases there was a thin film of something between the drops which they were unable to squeeze out, namely, water, paraffin, or air, as the case might be.

FIG. 56.

Will two soap-bubbles also when knocked together be unable to squeeze out the air between them? This you can try at home just as well as I can here, but I will perform the experiment at once. I have blown a pair of bubbles, and now when I hit them together they remain distinct and separate (Fig. 55).

I shall next place a bubble on a ring, which it is just too large to get through. In my hand I hold a ring, on

which I have a flat film, made by placing a bubble upon
it and breaking it on one side.   If I gently press the
bubble with the flat film, I can push it through the ring
to the other side (Fig. 56), and yet the two have not
really touched one another at all.   The bubble can
be pushed backwards and forwards in this way many
times.

I have now blown a bubble and hung it below a ring.
To this bubble I can hang another ring of thin wire,
which pulls it a little out of shape.   I push the end of

FIG. 57.   FIG. 58.

the pipe inside, and blow another bubble inside the first,
and let it go.   It falls gently until it rests upon the outer
bubble; not at the bottom, because the heavy ring keeps
that part out of reach, but along a circular line higher
up (Fig. 57).   I can now drain away the heavy drops of
liquid from below the bubbles with a pipe, and leave
them clean and smooth all over.   I can now pull the
lower ring down, squeezing the inner bubble into a
shape like an egg (Fig. 58), or swing it round and round,
and then with a little care peel away the ring from off
the bubble, and leave them both perfectly round every

way (Fig. 59). I can draw out the air from the outer bubble till you can hardly see between them, and then blow in, and the harder I blow, the more is it evident that the two bubbles are not touching at all; the inner one is now spinning round and round in the very centre of the large bubble, and finally, on breaking the outer one the inner floats away, none the worse for its very unusual treatment.

There is a pretty variation of the last experiment, which, however, requires that a little green dye called fluorescine, or better, uranine, should be dissolved in a separate dish of the soap-water. Then you can blow the

FIG. 59.

outer bubble with clean soap-water, and the inner one with the coloured water. Then if you look at the two bubbles by ordinary light, you will hardly notice any difference; but if you allow sunlight, or electric light from an arc lamp, better concentrated by the aid of a lens, to shine upon them, the inner one will appear a brilliant green, while the outer one will remain clear as before. They will not mix at all, showing that though the inner one is apparently resting against the outer one, there is in reality a thin cushion of air between.

Now you know that coal-gas is lighter than air, and so a soap-bubble blown with gas, when let go, floats up to the ceiling at once. I shall blow a bubble on a ring with coal-gas. It is soon evident that it is pulling upwards.

I shall go on feeding it with gas, and I want you to notice the very beautiful shapes that it takes. These are all exactly the curves that a water-drop assumes when hanging from a pipe, except that they are the other way up. The strength of the skin is now barely able to withstand the pull, and now the bubble breaks away just as the drop of water did.

I shall next place a bubble blown with air upon a ring, and blow inside it a bubble blown with a mixture of air and gas. It of course floats up and rests against the top of the outer bubble (Fig. 60). Now I shall let a little

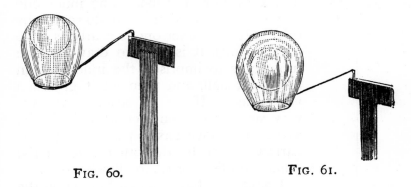

FIG. 60. FIG. 61.

gas into the outer one, until the surrounding gas is about as heavy as the inner bubble. It now no longer rests against the top, but floats about in the centre of the large bubble (Fig. 61), just as the drop of oil did in the mixture of alcohol and water. You can see that the inner bubble is really lighter than air, because if I break the outer one, the inner one rises rapidly to the ceiling.

Instead of blowing the first bubble on a heavy fixed ring, I shall now blow one on a light ring, made of very thin wire. This bubble contains only air. If I blow inside this a bubble with coal-gas, then the gas-bubble will try and rise, and will press against the top of the outer one with such force as to make it carry up the wire ring and a yard of cotton, and some paper to which the cotton is tied (Fig. 62); and all this time, though it is the inner one only which tends to rise, the two bubbles are not really touching one another at all.

A further variation of the experiment is illustrated in Fig. 63. The outer bubble is blown upon a wire ring held in the hand, and a smaller gas bubble is blown within it. Then with a second wire ring held in the other hand the bubble is touched so as to avoid the spot where the inner bubble rests against it, and then the outer one is drawn out into a cylindrical tube. The inner one may then be rolled up and down in the tube. If the inner one is nearly as large as the rings it is easy to draw out the tube so as to imprison the inner one in the lower half, and then let it up when you please. If the air space between is very small it can only crawl very slowly, as the air above has to pass through this narrow space to the other side of the inner bubble in order to allow it to rise.

I have now blown an air-bubble on the fixed ring, and pushed up inside it a wire with a ring on the end. I shall now blow another air-bubble on this inner ring. The next bubble that I shall blow is one containing gas, and this is inside the other two, and when let go it rests against the top of the second bubble. I next make the second bubble a little lighter by blowing a little gas into it, and then make the outer one larger with air. I can now peel off the inner ring and take it away, leaving the two inner bubbles free, inside the outer one (Fig. 64). And now the multiple reflections of the brilliant colours of the different bubbles from one to the other, set off by the

FIG. 62.

beautiful forms which the bubbles themselves assume,
give to the whole a degree of symmetry and splendour
which you may go far to see equalled in any other way.
I have only to blow a fourth bubble in *real* contact with
the outer bubble and the ring, to enable it to peel off
and float away with the other two inside.

We have seen that bubbles and drops behave in very
much the same way. Let us see if electricity will produce

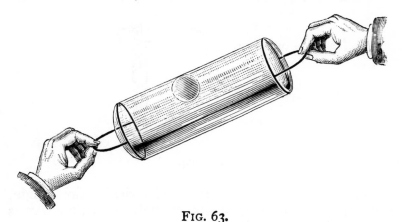

FIG. 63.

the same effect that it did on drops. You remember that
a piece of electrified sealing-wax prevented a fountain of
water from scattering, because where two drops met,
instead of bouncing, they joined together. Now there
are on these two rings bubbles which are just resting
against one another, but not really touching (Fig. 65).
The instant that I take out the sealing-wax you see they
join together and become one (Fig. 66). Two soap-
bubbles, therefore, enable us to detect electricity, even
when present in minute quantity, just as two water
fountains did.

We can use a pair of bubbles to prove the truth of one
of the well-known actions of electricity. Inside an electrical conductor it is impossible to feel any influence of

electricity outside, however much there may be, or however near you go to the surface.　Let us, therefore, take
the two bubbles shown in Fig. 57, and bring an electrified stick of sealing-wax near.　The outer bubble is a

FIG. 64.

conductor ; there is, therefore, no electrical action inside,
and this you can see because, though the sealing-wax is
so near the bubble that it pulls it all to one side, and

FIG. 65.

though the inner one is so close to the outer one that
you cannot see between them, yet the two bubbles
remain separate.　Had there been the slightest electrical
influence inside, even to a depth of a hundred-thousandth

of an inch, the two bubbles would have instantly come together.

There is one more experiment which is a combination of the last two, and it beautifully shows the difference between an inside and an outside bubble. I have now a plain bubble resting against the side of the pair that I have just been using. The instant that I take out the sealing-wax the two outer bubbles join, while the inner one un-

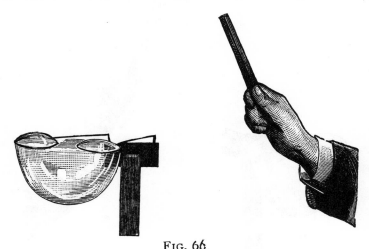

FIG. 66.

harmed and the heavy ring slide down to the bottom of the now single outer bubble (Fig. 67).

The Soap-bubble

It can only be our familiarity with soap-bubbles from our earliest recollections, causing us to accept their existence as a matter of course, that prevents most of us from being seriously puzzled as to why they can be blown at all. And yet it is far more difficult to realize that such things ought to be possible than it is to understand anything that I have put before you as to their actions or

their form. In the first place, when people realize
that the surface of a liquid is tense, that it acts
like a stretched skin, they may naturally think that a
soap-bubble can be blown because in the case of soap-
solution the "skin" is very strong. Now the fact is just
the opposite. Pure water, with which a bubble cannot be
blown in air and which will not even froth, has a "skin"

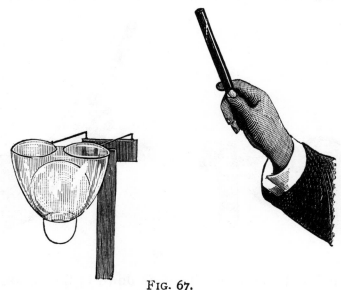

FIG. 67.

or surface tension three times as strong as soap-solution, as
tested in the usual ways, *e. g.* by the rise in a capillary
tube. Even with a minute amount of soap present the
surface tension falls off from about $3\frac{1}{4}$ grains to the linear
inch to $1\frac{1}{4}$ grains, as calculated from experiments with
bubbles by Plateau. The liquid rises but little more
than one-third of the height in a capillary tube. The
soap-film has two surfaces each with a strength or $1\frac{1}{4}$
grains to the inch and so pulls with a strength of $2\frac{1}{2}$
grains to the inch. Many liquids will froth that will not

blow bubbles. Lord Rayleigh has shown that a pure liquid will not froth though a mixture of two pure liquids, *e. g.* water and alcohol, will. Whatever the property is which enables a liquid to froth must be well developed for it to allow bubbles to be blown. I have repeatedly spoken of the tension of a soap-film as if it were constant, and so it is very nearly, and yet, as Prof. Willard Gibbs pointed out, it cannot be exactly so. For, consider any large bubble or, for convenience, a plane vertical film stretched in a wire ring. If the tension of two grains and a half to the inch were really identically the same in all parts, the middle parts of the film being pulled upwards and downwards to an equal extent by the rest of the film above and below it would in effect not be pulled by them at all and like other unsupported things would fall, starting like a stone with the acceleration due to gravity. Now the middle part of such a film does nothing of the kind. It appears to be at rest, and if there is any downward movement it is too slow to be noticed. The upper part therefore of the film must be more tightly stretched than the lower part, the difference being the weight of the intervening film. If the ring is turned over to invert the film then the conditions are reversed, and yet the middle part does not fall. The bubble therefore has the remarkable property within small limits of adjusting its tension to the load. Willard Gibbs put forward the view that this was due to the surface material not being identical with the liquid within the thickness of the film. That the surface was contaminated by material which lowered its surface tension and which by stretching of the film became diluted, making the film stronger, or by contraction became concentrated, making the film weaker. His own words are so apt and so much better than mine that I shall quote from his *Thermodynamics*, p. 313 : " For, in a thick film (as contrasted with a black film), the increase of tension with the extension, which is necessary for its stability with respect to extension, is connected with an excess of the soap (or of some of its components) at the surface as compared with the interior of the film."

This is analagous to the effect of oil on water described on p. 37. Lord Rayleigh has by a beautiful experiment supported the contamination theory, for he measured the surface tension of the surface of a soap-solution within the first hundreth of a second of its existence. He then

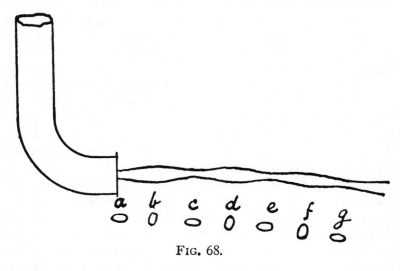

Fig. 68.

found it to be the same as that of water, for the surface contamination had no time to form. He allowed the liquid to issue from a small elliptical hole in a thin plate covering the end of a tube which came from a reservoir of the solution. When liquid issues from such a hole as at *a*, Fig. 68, the cross section of the stream being elliptical, as shown below, tends in virtue of the surface tension to beeome circular, but when it gets circular the movement alrcady set up in the section cannot be suddenly arrested and so the liquid continues its movement until it is elliptical in the other direction as at *b*, and this process is continued at a definite rate depending upon the surface tension and density of the liquid and size of the jet. At the same time the liquid is issuing at a definite rate depending on the depth of the orifice below the free surface, and so when the conditions are well chosen the liquid travels from *a* to *c* while the ellipse goes through its complete evolution, and this is repeated several times. Now if the surface tension were less the evolution would take longer and the distance between the nodes *a c e g* would be greater. With the same liquid head the distance between the nodes is the same with pure water and with soap-solution, showing that their surface tensions at first are the same, but with alcohol, which has its own surface tension from the beginning, the distance between the nodes is greater, as the surface tension is lower in a higher proportion than that in which the density is lower. Professor Donnan has quite recently shown by direct experiment of surpassing delicacy, that there is a surface concentration of the kind and amount required by Gibbs's theory.

The following experiment also indicates the existence of a surface concentration. If a soap-bubble is blown on a horizontal ring so that the diameter of the ring is very little less than that of the bubble, and the wetted stopper from a bottle of ammonia is brought close to the upper side of the bubble, it will immediately shrink away from the stopper and slip through the ring as though annoyed by the smell of the ammonia. Or, if below, it will retire to the upper side of the ring if the stopper is held below it. What really happens is that the ammonia combines with some of the constituents of the soap which are concentrated on the surface, and so raises the tension of the film on one side of the ring; it therefore contracts and blows out the film on the other side which has not yet been influenced by the ammonia. That part of the film influenced by the ammonia also becomes thicker and the rest thinner, as shown by the colours, which are then far more brilliant and variegated.

Going back now to the soap-film we see then that whatever its shape the upper parts are somewhat more tightly stretched than the lower parts, and in the case of a vertical film the difference is equal to that which will support the intervening film. There is, however, a limit beyond which this process will not go; there is a limit to the size of a soap-bubble. I do not know what this is. I have blown spherical bubbles up to two feet in diameter, and others no doubt have blown bubbles larger still. I have also taken a piece of thin string ten feet long and tied it into a loop after wetting it with soap solution and letting it untwist. Holding a finger of each hand in the loop and immersing it in soap-solution I have drawn it out and pulled it tight, forming a film in this way five feet long. On holding the loop vertical the film remained unbroken, showing that five feet is less than the limit with even a moderately thick bubble. With a thin bubble the limit should be greater still. Judging by the colour and by using the information given on the coloured plate which is explained in a subsequent chapter, the average thickness of the film must have been about thirty millionths of an inch and its weight about $\frac{8}{1000}$ of a grain per square inch. Taking a film an inch wide, the five feet or sixty inches would weigh close upon half a grain, that is about one-fifth of the total load that the film can carry, showing at least a 20 % capacity in the soap-film for adjusting its strength to necessity, which is far more than could have been expected.

I have also found that with the application of increased forces the bubble rapidly thins to a straw colour or white, so that the 20 % increase of load is not exceeded, but a film of this colour might be thirty-three feet high, or a black film ten times as much.

The feeble tension of $2\frac{1}{2}$ grains to the inch in a soap-film is quite enough in the case of the five-foot loop to make it require some exertion to keep it pulled so that the two threads are not much nearer in the middle than at the ends. In fact, this experiment provides the means by which the feeble tension of $2\frac{1}{2}$ grains to the inch may be measured by means of a seven-pound weight. If for instance the threads are seventy inches long, equidistant at their ends, and are $\frac{1}{16}$ inch nearer in the middle than at their ends when placed horizontally and stretched with seven-pound weights, then the tension of the film works out at $2\frac{1}{2}$ grains to the inch exactly. This is obtained as follows : The diameter of the circle of which the curved thread is a part is equal to half the length of the thread multiplied by itself and divided by the deviation of its middle point. The diameter of the circle then is $35 \times 35 \times 32$ inches. The tension of the thread, made equal to 7 lbs. or 7×7000 grains, is equal to the tension of the film in grains per inch multiplied by half the diameter of the circle of curvature of the thread. In other words the tension of the film in grains per inch is equal to the tension of the thread divided by half the diameter of the circle of curvature of the thread. The film tension then is equal to $\dfrac{7 \times 7000}{35 \times 35 \times 16}$ grains per inch. The fraction cancels at once to $2\frac{1}{2}$ grains per inch. As the upper parts of a vertical film are stretched more tightly than the lower parts, the pair of threads will be drawn together in the middle, rather more than $\frac{1}{16}$ inch, as the pair of threads are gradually tilted from the horizontal towards a vertical position, and then as the film drains into the threads and becomes lighter the threads will separate slightly.

Large bubbles are short-lived not only because, if it were a matter of chance, a bubble a foot in diameter with a surface thirty-six times as great as that of one two inches in diameter would be thirty-six times as likely to break if all the film were equally tender, but because the upper parts being in a state of greater tension have less margin of safety than the lower parts. These large bubbles however by no means necessarily break at the top. When a large, free-floating bubble is seen in bright sunlight on a dark background it is almost possible to follow the process of breaking. What is really seen, however, is a shower of spray moving in the opposite direction to that in which the hole is first made, while the air which cannot be seen blows out in the opposite direction to that of the spray. By the time it is done, and it does not take long, the momentum given to the moving drops in one direction and to the moving air in the opposite direction are equal to one another, but as the air weighs far more than the water the spray is thrown the more rapidly.

The breaking of a bubble is itself an interesting study. Duprée long ago showed that the whole of the work done in extending a bubble is to be found in the velocity given to the spray as it breaks, and thence he deduced a speed of breaking of 105 feet a second or seventy-two miles an hour for a thin bubble, and Lord Rayleigh for a thicker bubble found a speed of forty-eight feet a second or thirty-three miles an hour. The speed of breaking of a soap-bubble is curious in that it does not get up speed and keep going faster all the time, as most mechanical things do, it starts full speed at once, and its speed only changes in accordance with its thickness, the thick parts breaking more slowly. It may be worth while to show how the speed is arrived at theoretically. Take in imagination a film between two parallel and wetted wires an inch apart and extend it by drawing a wetted edge of card, india-rubber or celluloid along the wires. Then the work done in extending it for a foot, say, not counting of course the friction of the moving edge, is with a tension of $2\frac{1}{2}$ grains to the inch equal to $2\frac{1}{2} \times 12$ inch-grains, or if it is extended a yard to $2\frac{1}{2} \times 36$ inch-grains. Let us pull it out to such a length that the weight of the film drawn out is itself equal to $2\frac{1}{2}$ grains, that is to the weight its own tension will just carry. By way of example let us consider a film not very thick or very thin, but of the well-defined apple-green colour. This, according to the coloured plate is just under twenty millionths of an inch thick, and it weighs $\frac{5}{1000}$ or $\frac{1}{200}$ of a grain to the square inch. The length of this film one inch wide that will weigh $2\frac{1}{2}$ grains is therefore 500 inches or forty-two feet nearly. The work done in stretching a film of this area is $2\frac{1}{2} \times 500$ inch-grains. The work contained in the flying spray must also be $2\frac{1}{2} \times 500$ inch-grains. Now the work contained in a thing moving at any speed is the same as that which would be needed to lift it to such a height that it would if falling without obstruction acquire that speed. The work done in lifting the $2\frac{1}{2}$ grains through 500 inches is exactly the same as that done in pulling out the film 500 inches horizontally against its own tension, as the force and the distance are the same in both cases. The velocity of the spray, and therefore of the edge from which it is scattered is the same as that of a stone falling through a distance equal to the length of the film, the weight of which is equal to the tension at its end. Completing the figures a stone in this latitude falling forty-two feet acquires a velocity of fifty-two feet a second, which therefore is the speed of the breaking edge of the film. I conclude therefore, as the speed found in this example is intermediate between those found by Duprée and Lord Rayleigh, that I have chosen a film intermediate in thickness between those chosen by these philosophers. I would only add that a bubble has to be reduced to one quarter of its thickness to make it break twice as fast, then the corresponding length will be four times as great, and it requires a fall from four times the height to acquire twice the speed. A black film is about one thirty-sixth of the thickness of the apple-green film, it should therefore break six times as fast or 312 feet a second, or 212 miles an hour. The extra black film of half the thickness should break at the enormous speed of 300 miles an hour. These speeds would hardly be realized in practice as the viscosity of the liquid would reduce them.

Lord Rayleigh photographed a breaking soap-film by

placing a ring on which it was stretched in an inclined
position, and then dropping through it a shot wet with
alcohol, and about a thousandth of a second later
photographing it. For this purpose he arranged two
electro magnets, one to drop the wet shot and the other
to drop another shot at the same instant. The second
shot was allowed a slightly longer fall, so as to take a
thousandth of a second, or whatever interval he wanted,
longer than the first; it then by passing between two knobs
in the circuit of a charged Leyden jar let this off, and the
electric spark provided the light and the sufficiently short
exposure to give a good and sharp photograph. The
sharp retreating edge is seen with minute droplets either
just detached or leaving the film. Following Lord
Rayleigh I photographed a breaking film by piercing it
with a minute electric spark between two needles, one
on either side of the film, and by means of a piano-wire
spring like a mousetrap determining the existence of this
spark, and then a ten-thousandth of a second or more
later letting off the spark by the light of which it was
taken. My electrical arrangements were akin to those
by which I photographed bullets in their flight in one
thirteenth-millionth of a second, but my optical arrangements were similar to Lord Rayleigh's. One photograph
of a vertical film which had become a great deal thinner
in its upper part, is interesting as whereas all the lower
part has a circular outline, the upper part breaks into a
bay showing the speed of breaking upwards in the thin
film to be much greater. In all the photographs the
needle spark appears by its own light, so the point at
which the break occurs is seen as well as the circular
retreating edge.

Bubbles other than Soap-bubbles

While a solution of soap in soft water or water and glycerine is the most perfect material known for blowing bubbles, perfect by reason both of the ease with which it may be obtained and of its transparency and fluidity, yet some other materials allow of bubbles being blown. The best-known of these is melted resin to which should be added a small proportion, say one-tenth or one-twentieth, of beeswax or of gutta-percha or a smaller proportion of linseed oil. Whatever the mixture is it should be melted and well mixed and then bubbles may be blown. I have blown these with coal gas which then rise and remain adherent to the ceiling, but in a day or so they generally go to powder, and in any case they make too much mess to be tolerable in a living-room. Plateau, using a mixture of five parts of colophane and one of gutta-percha melted together at 300° F., obtained more permanent results. He was able to dip the wire frames described on p. 90 in such a mixture, even a cube as much as two inches in the side, and keep the exquisite film structure developed within on removal for two years.

Surprising and ludicrous bubbles may be blown with a solution of saponine. Saponine may be bought as a white powder and very little dissolved in water will give the required mixture. A sufficiently good solution may be obtained by cutting up horse chestnuts in thin slices and soaking them in very little water. The slightly yellow liquid, which contains other things as well, contains enough saponine to enable bubbles up to three or four inches in diameter to be blown. With either of these solutions moderate-sized bubbles may be blown, but it is well to have a pipe with the air-way very much constricted so that the blowing shall not be too rapid. These bubbles when being blown or when contracting slowly under their own tension and blowing air back through the pipe show nothing out of the common. They seem weak and tender, that is all. But, having a bubble an inch or more in diameter, suck a little air out through the pipe when at once a marked peculiarity is

FIG. 69.

noticeable. The bubble cannot contract and follow the air as a soap-bubble does, but forms a wrinkled bag (Fig. 69) which if left will slowly regain its spherical form, or which will do so at once under compulsion if air is blown in again. This may be repeated many times and the result, especially when magnified and projected on a screen, is quite ludicrous.   If sufficient air is drawn out the puckered bag becomes sharp pointed and yet it blows out into the spherical form again.   The peculiarity of a saponine solution is the rigidity of its surface while its interior is fluid.   Plateau made an elaborate investigation of this property of some fluids which is more marked in saponine than in any other.   Saponine-bubbles are so flimsy and tender compared with soap-bubbles that the idea may be formed that the tension is less.   The opposite is really the case.

Lord Rayleigh has found that if two equal bubbles, one of soap and one of saponine, are blown on two pipes and the two pipes are connected together, then the saponine-bubble contracts and blows air into the soap-bubble, showing the tension of the saponine-bubble to be the greater. He found that the soap-bubble had to be reduced to about two-thirds of the diameter of the saponine-bubble for the balance which, as we have seen, is an unstable one to be obtained, showing the tension of the soap-film to be about two-thirds of that blown with saponine.   I have found that the saponine-solution is quite admirable in the froth apparatus (Fig. 52).   For this purpose a solution of saponine in 1000 times its weight of water does very well.   On making the froth the same structure is seen as with a soap-froth, but within the first few seconds a marked difference appears especially if a good magnifying lens be used.   Every rectangular element of film is seen to contain a colour pattern parallel to its periphery showing even better than soap the action of the suction described by Willard Gibbs (see p. 162).   As the films are so rigid light patches of colour are unable to move upwards through them but remain where formed, giving rise to the rectangular patterns.   Then, when a film joining others breaks, it leaves its mark behind in the remaining film which thus carries its history written in white areas and coloured patterns, as the transverse ridges on a person's teeth tell the history of early disturbances in nutrition. Then when a film breaks, especially if a longish piece, the process can be watched, and the jerky retreat of the edge can be followed by the eye in marked contrast to the great speed of rupture of the highly liquid soap-solution.

The addition of glycerine to the saponine-solution up to half the original volume, while leaving the formation of concentric colour patterns much the same as with the simple solution in water, makes all the movements still more slow, and the crawling of the films to new positions of equilibrium after one is broken or the retreat of a broken edge are now so sluggish that they may be watched by the naked eye. A single drop of soap solution in an ounce of saponine-solution completely destroys the peculiar property of saponine which prevents the film from contracting quickly. With another drop or two the solution behaves as a soap-solution and the film is freely fluid. Saponine present even in very small quantity will make water froth ; one part in 100,000 is evident. Saponine powder is very light and may easily be inhaled accidentally, it is then most unpleasant, producing an irritation of the nose and throat. The smell and taste are unpleasant. Its property of making a persistent and clammy froth is useful in making up artificial drinks for people who are afraid of wine or beer.

Bubbles of pure mercury in water have already been described, p. 67. These should be referred to here as an example of a beautiful bubble not made of soap-solution.

Bubbles may of course be blown with melted glass and melted quartz, the latter at a temperature at which steel and fireclay are fluid as water, but the possibility of blowing such bubbles of fairly uniform thickness depends on quite a different property from that described by Willard Gibbs, p. 105, and it is free from the surface viscosity of saponine. The reason that glass may be blown into such perfect forms is that it is viscous as treacle is viscous and it is more viscous as it gets cooler.

Now in blowing glass the thinner parts cool most quickly and become most viscous, and so the more fluid, thicker parts continue to get thinner while the thinner parts resist and this tends towards uniformity. Actually the skilled glassblower accentuates this property of glass by turning it round, first with a view to keep it all at one temperature, and then, when any part shows a disposition to get extended too much he turns this round so as to be at the lowest position when the rising current of air, which any hot object induces, strikes upon this part and cools it and prevents its further stretching. The glassblower by exercising his skill in blowing, waiting, turning and watching the glass and sometimes even using a fan, makes it obedient to his wishes and the process most fascinating to watch. Glass blown violently and unskillfully makes large, irregular-shaped bulbs which may easily become so thin as to show the colours of the soap-bubble. Very small beads of hot glass at the end of a narrow tube show this best. I believe that resin-bubbles can be blown for the same reason that glass and fused quartz can, because the thinner places are cooler and more viscous, rather than that there is any stability due to surface concentration of some constituent as in a soap-bubble.

White of egg and water stirred together froth well, but allow of bubbles being blown to a very limited extent. White of egg and gelatine or glue, which acts much in the same way, go bad on keeping. A minute trace of either however in water is valuable, as a glass plate wetted with this and left to dry will allow drawings, etc., to be made upon it in ink without running of the ink, and this is useful for extemporized lantern slides.

Permanent films so perfect as to show all the colours of soap-bubbles, and so thin as to show the thinnest white and even the black, may be made by dropping celluloid solution on to water. Mr. Glew and Mr. Thorp have both used celluloid films, the latter for his beautiful copies of diffraction gratings. Mr. Glew has written to me as follows: " I found celluloid dissolved in amyl acetate, 8 grains to the ounce (or 1 gramme in 14·6 cubic centimetres), gave the best results. . . . There is no difficulty in getting very clear films if the amyl acetate is quite free from moisture. To ensure this I boil the amyl acetate for a few minutes. (This liquid is very inflammable.) If there is a trace of moisture there is a rich bloom or want of clearness over the whole film. The drops of the solution should fall about $1\frac{1}{2}$ inches above the surface of the water, or the drops may roll about on the air-film and so delay spreading. Black films may be raised two or three inches in diameter. Coloured films, second and third order, any size, may be raised by the assistance of a net underneath the film. I have thus obtained some over a foot in diameter. The strength of these films is to me very extraordinary; one three inches in diameter will bear several ounces of sand if put on with care. The films are extremely elastic, and are thrown into nodes and loops by any sounds of short wave-length. These are best seen by sunlight reflected from the vibrating film on to a sheet of white card. Whistling or singing near the film gives a charming combination of motion and colour on the screen, and can be arranged for a lecture experiment of great beauty. Sand spread on the film gives a distinct pattern for each note. The sand pattern can be picked off by placing a sheet of gummed paper over and blowing gently to secure contact, and so pick up the sand or any powder, giving a permanent record. Films can be silvered, and so make very light mirrors on a quartz or glass frame. Two or three films superposed allow passage of alpha particles, but do not allow the gaseous emanations to pass. This is useful in certain research work. Lord Rayleigh has made use of these films for enclosing various gases for refraction of sound waves. As to durability, I have some which I made about twenty years ago and they show no change. If more widely known they would be useful in research work. A red-hot iron or flame may be held about half an inch above a coloured film without affecting it, radiation passing freely through the film, but the ions cannot pass through, which can be seen by bringing the knob of a charged electroscope under the film, which thoroughly protects it from discharge by ions." I have given part of Mr. Glew's letter verbatim as he has made such excellent use of these films, and was, I believe, the first to make them. I would only add, by way of further instruction, that the solution should be dropped upon the surface of clean cold water in a large basin, and left until the amyl acetate has evaporated and colours are showing. The film may then be lifted from the surface of the water by means of a wire ring, and then exposed to the air. It then dries more completely, and it becomes tense and very brilliant in its colours. If any parts are black they are, of course, so thin as to be very tender.

Prof. Wood recommends that thin sheet celluloid should be boiled in water half an hour before being dissolved in the acetate of amyl, to get rid of free camphor. Mr. Thorp requires for making his copies of diffraction gratings celluloid films of the greatest perfection. These gratings may have 15,000 or more parallel lines to the inch, and it is essential that each line should be in its proper place and that the whole film should be uniform. He has given me the following particulars of the methods he follows in making such films. One part of Schering's celloidin is dissolved in fifty parts of amyl acetate and is then allowed to settle, for three months if possible. It is then poured on to a piece of perfectly level and clean plate glass, and allowed to dry in a dust-proof box. A short piece of glass tube is then taken which has been ground at one end on a convex lens-maker's grinding tool and polished, and then on a concave lens-maker's grinding tool and polished, and then ground on a flat surface, so as to make a narrow flat between the two sloping faces, and then polished. A solution of seccotine is brushed round the outside of the tube, but not right up to the polished end, and this is allowed to get nearly dry or set. A circle is scribed round in the celluloid film, still on the plate glass, about one-fifth of an inch more in diameter than the outside of the glass tube, and the polished end of the tube is placed centrally upon it. The cut edge of the film is then breathed upon, and the point of a sharp knife inserted so as to raise the film and press it gently against the adhesive surface. This is done all round, and the whole is then left to become perfectly dry and hard. As the film is in optical contact with the supporting glass plate, it is necessary to run a little water all round the outer edge, and to breathe on the film within to loosen it. It can then be readily detached, and the resultant film when dry is a perfect plane film, showing by reflection interference colours. It may be made so thin that these are very brilliant.

Composite Bubbles

A single bubble floating in the air is spherical, and as we have seen this form is assumed because of all shapes that exist this one has the smallest surface in relation to its content, that is, there is so much air within, and the elastic soap-film, trying to become as small as possible, moulds the air to this shape. If the bubble were of any other shape the film could become of less surface still by becoming spherical. When however the bubble is not single, say two have been blown in real contact with one another, again the bubbles must together take such a form that the total surface of the two spherical segments and of the part common to both, which I shall call the interface, is the smallest possible surface which will contain the two quantities of air and keep them separate. As the soap-bubble provides such a simple and pleasing way of demonstrating the solution of this problem, which is really a mathematical problem, it will be worth while to devote a little time to its consideration. Let us suppose that the two bubbles which are joined by an interface are not equal, and that Fig. 70 represents a

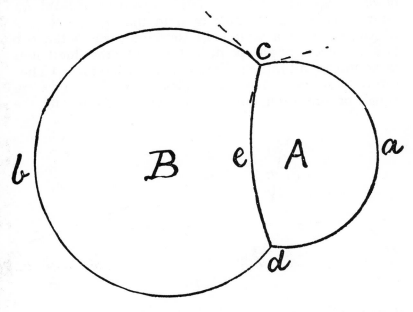

FIG. 70.

section through the centres of both, A being the smaller and B the bigger bubble. In the first place we have seen (p. 52) that the pressure within a bubble is proportional to its curvature or to 1 divided by the radius of the bubble. The pressure in A, by which I mean the excess over atmospheric pressure, will therefore be greater than the pressure in B in the proportion in which the radius of B is greater than the radius of A, and the air can only be prevented from blowing through by the curvature of the interface. In fact this curvature balances the difference of pressure. Another way of saying the same thing is this : the curved and stretched film *dac* pushes the air in *A* to the left, and it takes the two less curved but equally stretched films *dbc* and *dec*, pushing to the right to balance the action of the more curved film *dac*. Or, most shortly of all, the curvature of *dac* is equal to the sum of the curvatures *dbc* and *dec*. Now consider the point *c* or *d* in the figure either of which represent a section of the circle where the two bubbles meet ; at any point in this circle the three films meet and are all three pulling with the same force. They can only balance when the angles where they meet are equal or are each angles of 120°. Owing to the

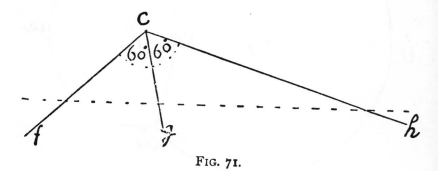

FIG. 71.

curvature of the lines, these angles do not look equal, but I have dotted in at *c* tangents to the three curves at the point *c*, and they clearly make equal angles with each other.

This equality of the angles is not an independent proposition to the last with regard to the curvatures ; if either condition is fulfilled the other necessarily follows, as also does the one I opened with that the total surface must be the least possible. Plateau, the blind Belgian professor, discussed this, as he did everything that was known about the soap-bubble, in his book *Statique des Liquides*, published in Brussels, a book which is a worthy monument of the brilliant author. He there describes a

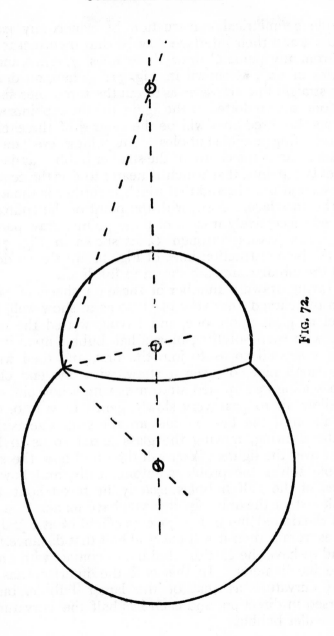

FIG. 72.

simple geometrical construction by which any pair of
bubbles and their interface may be drawn correctly.

From any point C draw three lines, cf, cg, ch, making
angles of 60°, as shown in Fig. 71. Then, on drawing
any straight line across so as to cut the three lines such as
the one shown dotted in the figure, the three points where
it cuts the three lines will be the centres of three circles
representing possible bubbles. The point where it cuts the
middle line is the centre of the smaller bubble, and of the
other two points, that which is nearer to C is the centre of
the second bubble, and that which is further is the centre
of the interface. Now, with the point of the compasses
placed successively at each of these points, draw portions
of circles passing through C, as shown in Fig. 72, in
which the construction lines of Fig. 71 are shown dotted
and the circular arcs are shown in full lines.

Having drawn a number of these on sheets of paper,
with the curved lines very black so as to show well, lay a
sheet of glass upon one, and having wetted the upper
side with soap solution, blow a half bubble upon it, and
then a second so as to join the first, and then with a
very small pipe, or even a straw, with one end closed
with sealing-wax opened afterwards with a hot pin so as
to allow air to pass very slowly, gently blow in or draw
out air until the two bubbles are the same size as those
in the drawing, moving the glass about so as to keep
them over the figure. You will then find how the soap-
bubble solves the problem automatically, and how the
edges of the half bubbles exactly fit throughout their
whole extent the drawings that you have made.

If the dotted line in Fig. 71 cuts cf and ch at equal distances from c, then it will cut cg at half that distance from
c, and we have the case of a bubble in contact with one of
twice its diameter. In this case the interface has the
same curvature as that of the larger bubble, but is
reversed in direction, and each has half the curvature of
the smaller bubble.

If the dotted line in Fig. 71 cuts *cf* and *cg* at equal distances from *c*, it will be parallel to *ch* and will never cut it. The two bubbles will then be equal, and the interface will then have no curvature, or, in other words, it will be perfectly flat, and the line *cd*, Fig. 70, which is its section, will be a straight line.

There are other cases where the same reciprocal laws apply as well as this one of the radii of curvature of joined soap-bubbles. It may be written in a short form as follows :

$$\frac{1}{A} = \frac{1}{B} + \frac{1}{e},$$ using the letters of Fig. 70 to represent the

lengths of the radii of the corresponding circles. For instance, a lens or mirror in optics has what is called a principal focus, *i.e.* a distance, say A, from it at which the rays of the sun come to a focus and make it into a burning-glass. If, instead of the sun, a candle flame is placed a little way off, at a distance, say B, greater than A, then the lens or mirror will produce an image if the flame at a distance,

say *e*, such that $\frac{1}{A} = \frac{1}{B} + \frac{1}{e}.$ Or, again, if the electrical

resistance of a length of wire, say A inches long, is so much, then the electrical resistance of two pieces of the same wire, B and e inches long, joined so that the current is divided between them, will be the same as that of A if

$$\frac{1}{A} = \frac{1}{B} + \frac{1}{e}.$$

The soap-bubble then may be used to give a numerical solution of an optical and of an electrical problem.

Plateau gives one other geometrical illustration, the proof of which, however, is rather long and difficult, but which is so elegant that I cannot refrain from at least stating it. When three bubbles are in contact with one another, as shown in Fig. 73, there are of course three interfaces meeting one another, as well as the three bubbles all at angles of 120°. The centres of curvature

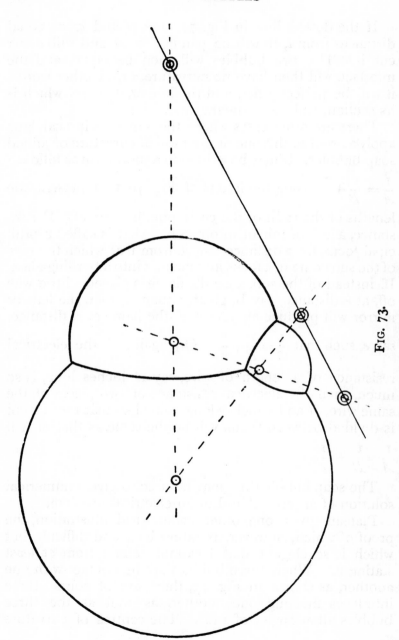

FIG. 73.

also of the three bubbles and of the three interfaces, also necessarily lie in a plane, but what is not evident and yet is true is that the centres of curvature, marked by small double circles of the three interfaces, lie in a straight line. If any of you are adepts in geometry, whether Euclidian or analytical, this will be a nice problem for you to solve, as also that the surface of the three bubbles and of the three interfaces is the least possible that will confine and separate the three quantities of air. The proof that the three films drawn according to the construction of Fig. 71 have the curvatures stated is much more easy, and I should recommend you to start on this first. If you want a clue, draw a line from the point where the dotted line cuts *cg* parallel to *cf*, and then consider what is before you.

Out-of-door Bubbles

Beautiful as the soap-bubble is as an object indoors it is still more so out of doors, especially when large with the sky reflected in duplicate from its upper and lower coloured faces. I strongly commend these experiments to people who are seeking for a pleasing occupation in the garden for a summer afternoon. The first thing you will notice is the curious shape of the skyline showing in spherical perspective the surrounding trees or buildings. Fig. 74 shows in spherical perspective the back of a group of offices in Victoria Street. This is a dismal picture to take as an example. I would, however, suggest that those who have the advantage of beautiful surroundings might find it interesting to photograph them as presented in a bubble. In this picture the buildings behind the camera are seen in the middle, those at the side curiously bent and distorted, while the high block of buildings with its windows directly in front is seen upside down at the top of the bubble. Further the whole thing is repeated in a reversed position below, and there is some confusion from the two pictures overlapping. I cannot refrain from including here a companion picture

FIG. 74.

(Fig. 75) taken indoors, showing a kind of dreamland portrait taken by reflection from the evanescent surfaces of a simple soap-bubble. By this means any number may be included in a group vanishing in the smallness of distance around a central figure, and, of course, if the inverted repetition of the whole picture is not wanted a half bubble only may be used. I shall leave to the reader to find out how to arrange the bubble, camera and surroundings as a simple problem which may be interesting to those with photographic inclination. You will not see

FIG. 75.

great and conspicuous windows reflected in out-of-door bubbles, though you may in pictures of them, but you will see great sheets of brilliant and varying colour. While out-of-door bubbles are so attractive it is best not to blow them out of doors for two reasons. It is seldom that there is so little wind that bubbles of any size can be

managed, and as it is far better that they should be
blown with a very little gas as well as air so that they
tend neither to rise nor fall but follow like thistledown
every movement of the air, it is more convenient on this
account also to blow them within easy reach of a gas-jet,
unless, as is not likely, you have a portable gasholder
or gas-bag such as was used before the days of steel
cylinders for limelight purposes.   Actually one of these
cylinders charged with coal gas or with hydrogen will
do, but the gas pressure even with a regulator is such
that the manipulation is not so easy, and, of course, with
hydrogen, which has about twice the lifting power that
coal-gas has, less gas will be required.   For these reasons
I shall describe what I have found to be the easier
way.   Choose a day on which there is a bright sun
or white clouds and but little wind, and a window
out of which the air from the house is gently moving.
This may be regulated to a certain extent by opening or
closing other windows or doors so as to control the
inlet of air to the house on the windward side, or, if the
outdraught is too much, by opening other windows on
the leeward side.   There should be within range of an
india-rubber tube a gas-jet to which one end of the
tube should be attached, and to prevent the tube from
kinking it should be tied about a foot from the end to the
jet so that the intervening piece makes an easy bend, and
the length of tube leaves the place where it is tied in the
direction required.   It is convenient also to have a
ticking clock near enough to be heard.

Having all these things duly arranged, and a pipe and
soap solution, preferably that described on p. 170, hold
the end of the india-rubber tube with one hand with which
also it is pinched to prevent the gas escaping, dip the
end of the india-rubber tube into the solution, and then
on taking it out let the gas blow a bubble.   When the
bubble is the size of an apple or an orange, the size
depending on the weight of liquid in the bubble, the

lifting power of the gas will just carry it, so that when detached it will neither rise nor fall. If the end of the india-rubber tube is small the amount of liquid will be small, and quite a small bubble will just float; if the tube is connected to a pipe with a larger end the weight will be more, and the bubble that will just float will be larger. It is easy having the gas-tap turned to any particular position to ascertain how much gas will be about right by counting the ticks of the clock while the gas is flowing and then detaching the bubble. With fairly uniform manipulation the number of ticks found by experience is a sufficient guide. I have shown on p. 134 how you can tell by calculation what size of bubble will just float, but with the instructions here given there is no difficulty in succeeding without any calculation. When a bubble has been blown so as just to float, and this is wafted out of the window, it will float away, and if it is in a town among buildings, and not among trees, it will last a sur-prisingly long time, being blown hither and thither, skirt-ing walls, but not touching them, and withstanding in an amazing way the action of eddies; now resting stationary or nearly so, now being buffeted by a whirl of air and spun round, or drawn out of shape like a sausage, or even, as I have seen, separated into two or three distinct bubbles. It is surprising, too, how far bubbles of moderate size will travel in the rain. Among trees, however, their life is short, they move with the air into the branches, and all is over. Bubbles blown with coal gas alone just large enough to float are too small to produce the great areas of varied colour which are so striking with large bubbles, but more gas must not be blown in, for if it is, the bubble will rise and disappear into the sky. However, as the soap solution and the gas balance each other, air may be blown in without destroying the balance, as the weight of air blown in is just balanced by the buoyancy of the outer air. It is therefore only neces-sary in order to blow larger bubbles that will just float to blow from a wide-mouthed pipe a bubble with air, and while blowing it to insert into its side the wetted end of the india-rubber tube connected with the gas supply, and allow the gas to enter for the number of ticks of the clock that will be found suitable. The time may not be the same as before, but more if the pipe used picks up a greater weight of liquid. Sometimes it is convenient to support the bubble on a ring of wire with a handle, or with three feet so as to stand as a tripod, the ring being two or three inches in diameter, and, of course, previously wetted with the solution. The chief objection to this is the difficulty of letting the bubble go without breaking it. The right movement to give to the ring is a very gentle movement straight away from the bubble, and not sidelong, but practice will show better than any description. Another way of causing a large bubble to leave a ring without shock is to blow a second smaller bubble in actual contact with the first bubble and the ring. The smaller bubble may then be made to enter the ring, when the larger bubble will draw off with but little disturbance. Single bubbles a foot in diameter or so may thus be blown and detached, and though such large bubbles do not last so long as small ones, they last quite long enough to present a most beautiful sight as they sail across a garden.

When the amount of gas is not well chosen the bubbles of course slowly rise or sink as the case may be. Occasionally when the gas was just insufficient I have seen a large bubble move slowly over a paved area, and there wipe off the pendent drop which generally drains to the bottom, and then relieved of this load it slowly rose again and sailed away. Oftentimes as they sail a drop falls away, and then the change of their course due to loss of weight becomes manifest.

I have already explained how the interfaces of multiple bubbles may be made either plane or very slightly curved by making the adjacent bubbles equal or nearly so. Groups of equal or nearly equal bubbles three or four inches in diameter, and grouped in twos, threes, fours, or fives may be blown by the aid of a supporting ring, and, if desired, the coal gas to support the combination may be put into one only, which, of course, will then carry the rest below it. Such multiple bubbles are even more magnificent than single bubbles, because the flashing of the sunlight reflected from the large plane faces in tints of ruby, emerald and sapphire contrasts with the steady and more restricted spots of light reflected from the spherical surfaces.

A further variety of multiple bubble which may be described as a magnified froth is interesting but not so beautiful as the groups just described. It is, however, very easily made. Place a saucer in a basin and pour in some soap solution, dip the end of the india-rubber pipe below the surface and fix it there, and regulate the flow of gas so that the bubbles that rise to the surface may be rather less than an inch in diameter. Such a single bubble unless it is very thin will not rise even if free, and it certainly cannot pull itself away from the liquid. As, however, the bubbles are being produced at a great rate a mass of froth of a coarse texture is formed, and the liquid drains from the upper bubbles until they are light enough to support themselves. The result is that at first a great hemispherical mass of froth is formed, and then as the upper part gets lighter it begins to pull upwards and the mass elongates until a column perhaps a foot high and of about the same diameter as the saucer gradually rears up and then pulls itself off and sails away. Such a mass may rub against a wall as it goes, destroying the surface bubbles in the process and then sail on rather smaller than before. The reason for putting the saucer in a basin is that the mass of froth carries much of the liquid over the edge which would otherwise be wasted. If the mass of froth clings to the basin it may not detach itself so readily, and in that case it may be well to stand the saucer on a mug in the basin. Of course these masses of froth illustrate very well the general characteristic of froth described on p. 92, of having never more than three faces meeting in a line, or more than four in a point, and always so meeting as to make angles of 120°. Groups of four or five large bubbles of equal size give a perfect representation of the rhombohedral end of the cell of a honeycomb.

It may be worth while to see how much gas is needed to support a bubble of any colour, or how small a bubble of any colour can be supported by coal gas. The following figures will make it easy to find out: $3\frac{1}{4}$ cubic inches of air weigh one grain. Coal gas is about $\frac{2}{5}$ the weight of air, that is if it does not contain too much water gas, so its lifting power is about $\frac{3}{5}$ of the weight of its volume of air. One cubic inch will therefore lift $\frac{3}{5} \times \frac{4}{13}$ grain. This is $\frac{12}{65}$ or more conveniently $\frac{185}{1000}$ grain. A spherical bubble contains slightly over half as many cubic inches as the cubical box it will just enter contains. The actual factor to convert the volume of a cube into that of the sphere of the same diameter is ·5236, which is a little over one half; or, the volume of a sphere is exactly $\frac{2}{3}$ of the volume of the containing cylinder. The surface of a sphere is exactly equal to that of the curved surface of the containing cylinder, and that is ·7854 or a little over three-quarters of the surface of four faces of the containing cube. To take an example let the bubble have a diameter of one inch. The volume of the containing cube is one cubic inch, and therefore the one-inch sphere contains ·5236 cubic inch. The surface of four faces of a one-inch cube is 4 square inches, and this multiplied by ·7854 gives 3·1416 square inches, and this is the surface of the one-inch sphere. The weight of this in thousandths of a grain per square inch for a soap-film of any colour can be read off directly from the scales on the coloured plate. Again, by way of example take the bright apple-green weighing $\frac{5}{1000}$ grain for every square inch.

Then our green one-inch bubble will contain $\dfrac{5 \times 3\cdot1416}{1000}$

grain of water, that is $\dfrac{15\cdot708}{1000}$ grain. If filled with coal

gas the lifting power of the gas will be $\dfrac{\cdot5236 \times 185}{1000}$, or

$\dfrac{96\cdot7}{1000}$ grain. This is a little more than six times as much
as the bubble weighs. If therefore the green bubble
is made only one-sixth of an inch in diameter instead of
one inch the surface and therefore the weight will be
reduced in the ratio of $6 \times 6 : 1$, while the quantity of
gas and therefore its lifting power will be reduced in the
ratio of $6 \times 6 \times 6$ to 1. A bubble one-sixth of an inch in
diameter of the bright apple-green colour filled with coal
gas will therefore just float in the air. The scale on the
coloured plate shows that the thinnest white film weighs
just about one thousandth part of a grain to the square
inch, or one-fifth as much as the green film. The smallest
white film that will just float when filled with coal gas
is therefore one-thirtieth of an inch in diameter, or
the size of a small pin's head. Very small and very thin
bubbles are difficult to blow even with very small pipes,
but bubbles such as I have been describing are often
formed unintentionally when a single bubble is pulled
into two, and I have occasionally seen these minute
and very thin bubbles floating in the air. It may be
interesting to know that the same scale which gives the
weight of films in grains per square inch may also be
used to find the weight in centigrammes per square centimetre, as the two systems of measurement differ by less
than one half of one per cent. So of course the scale
that gives thousandths of a grain per square inch gives
also thousandths of a centigramme, or hundredths of a
milligramme per square centimetre. It also gives the
thickness in ten thousandths of a millimetre.

The Colour and Thickness of Bubbles

The colours of soap-bubbles are so beautiful and varied that that alone is sufficient reason for wishing to know something as to the cause of the colour. As the colour depends upon the thickness primarily, and it is convenient at times to know the thickness and hence the weight of a bubble, this is an additional reason for trying to understand how it is that at some times they are so brilliantly coloured, while at other times they may have no colour at all, and how when they are adequately protected they may even be almost perfectly black. However, it is not easy to know how to begin, as the explanation requires a formidable chapter of optics, and it is not worth giving at all unless it is done with sufficient completeness to lead up from the simplest first principles to the observed result.

Light has been found by experiment to travel at the rate of very nearly a thousand million feet per second, or 186,680 miles per second, or seven and a half times round the earth per second, or thirty thousand million centimetres per second, that is in a vacuum; and very nearly or almost exactly at the same rate in air. In a vacuum, light of any colour travels at the same rate, and in air almost exactly so. In water, however, light travels three-quarters of the rate that it does in air, and lights of different colours do not all travel at exactly the same speed; but red is a little faster than yellow, and this than green, and blue and violet fall off a little more still. As the very slight differences in the rate at which lights of different colour travel in water are of quite secondary importance so far as the colours of soap-bubbles are concerned, I shall not refer to this again, but take light as travelling three-quarters as fast in water as it does in air. It is almost exactly three-quarters for all the colours.

Now suppose light to enter a trough of water obliquely

as indicated by the arrow 1 in Fig. 76, and that a beam
of light between the lines *ab* and *cd* only is being considered.   Light which left the sun, say at a particular
moment, will at some time reach the water at *b*.   At this
exact moment light which left the sun at the same time
that that at *b* did but travelling along *cd* will have reached

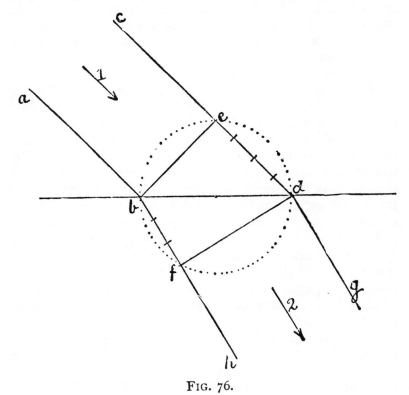

FIG. 76.

the point *e*, and will still have all the way from *e* to *d* to
go before it reaches the water.   The angle *bed* of course
is a right angle, and the point *e* may be found by drawing
a semi-circle shown dotted on *bd*.   While the light is
going from *e* to *d* the light which had reached *b* will travel
three-quarters of the distance *ed* in the water.   Set the

compasses to a radius equal to three-quarters of *ed*, and with one point at *b* mark the point *f* on the lower semicircle. The light leaving *b* will therefore travel as far as *f* while the light leaving *e* is getting to *d*. Join *fd*, then light which at one moment had reached the line *be* will later have reached the line *fd*, and *dg*, *fh* at right angles to *fd* will now be the boundaries of the light in the water, and the light will be travelling in the direction of the arrow 2. Light then is bent in its course when entering water obliquely, and the construction of Fig 76 shows how much and why.

If the water is contained between two parallel walls, as in a soap-bubble, then a repetition of the construction on the side where the light leaves the water will show that the beam of light there becomes more inclined again than it was in the water, and that it comes out parallel to the direction of the arrow 1 by which it entered, but the beam is displaced laterally by a small proportion of the thickness of the soap-bubble, and that you do not notice when you look at anything through a soap film.

If instead of water, through which light travels three-quarters as fast as it does through air, glass is taken, through which light travels two-thirds as fast as it does through air, then, following the construction of Fig. 76, it is only necessary to make *bf* two-thirds instead of three-quarters of *ed*.

This bending of light as it enters a dense medium is called refraction, and the construction of Fig. 76 is an illustration of the well-known optical "law of sines." It is necessary to understand it in order to follow the course of light in the thickness of a soap-bubble when tracing the cause of the brilliant colours.

If you take the trouble to make this construction for a variety of inclinations of *ab* from the perpendicular to nearly the horizontal, you will find that with the steeper angles *bf* is inclined from the perpendicular almost exactly three-quarters as much as *ab* is, but that as *ab* becomes more nearly horizontal *bf* changes in inclination very little; and further, that when *ab* is as low as it can be, so that *ed* and *bd* are the same, *bf* becomes three-quarters of the diameter, and this corresponds to an inclination within the film of 41°. This is the least inclined direction in which light can enter water from air, and similarly it is the smallest inclination at which a fish, say, can see out of the water when looking at things just above the water level. Prof. Wood has taken photographs with an immersed camera showing the kind of distorted view of the surrounding landscape that a fish can see. If light within water meets the surface at an angle less than 41° the construction of Fig. 76 becomes impossible, and there is no escape for the light, which therefore becomes totally reflected back into the water.

The next peculiarity of light, and one much more difficult to understand is the fact that, unlike a stream of water which pours continuously forward, light is a particular kind of interrelated electrical and magnetic disturbance, which for instance for a particular green of the spectrum repeats itself every one fifty-thousandth of an inch ($\frac{1}{50000}$ inch), that is, the wave-length is twenty-millionths of an inch long in air. In this fifty-thousandth of an inch the electrical and magnetic conditions become reversed in the first hundred-thousandth of an inch, and then reversed again in the second hundred-thousandth of an inch; and this process is repeated every fifty-thousandth of an inch in a vacuum or in air, while the light is travelling at the rate of nearly 12,000,000,000 inches a second. It will be seen then that with the particular green light the electrical and magnetic cycles occur at the rate of 50,000 × 12,000,000,000 times a second, or 600,000,000,000,000,000 times a second. If light were like a continuous stream of water it would not be possible to take two streams and add them together and find they produced nothing at all—that they seemed, in fact, to be wholly destroyed. As, however, the electric and magnetic causes of green light are reversed in direction every hundred-thousandth of an inch, it is only necessary to do something to split up a beam of this green light and shift one part a hundred-thousandth of an inch backwards or forwards with respect to the other half, keeping them in the same path, for the North magnetic action of one to neutralize the South magnetic action of the other, and the positive electricity of the one to neutralize the negative electricity of the other, and for the result to be darkness. This, at first, seems incredible. If one candle makes a certain light in a room two candles make twice as much, and no one ever saw a room suddenly become dark when a second candle was lighted. For the action that I have described to take place two separate sources of light must not be used, nor must two separate beams of light from the same source, but a single beam of light must be used ; then and then only can the production of darkness by its two parts be brought about. This action is called interference, and examples of interference are found in other rhythmical operations as well as in light. For instance, in a twin-screw steamer, if one engine is going very little faster than another the vibrations due to the two engines alternately act together and make a greater effect than either, and in opposite directions, so that the result is less than that due to either separately. Or if a tuning-fork that has been struck is held near the ear and slowly turned round, four places will be found where the actions of the two prongs are equal and opposite, and in these no sound or but very little is heard.

The figures that I have given for the wave-length and number of cycles in a second seem sufficiently astounding, but when once the imagination gives up the contest and leaves it to reason, the difficulties are not so great, for there is no very great difficulty in the experimental measurement of either velocity or the wave-length, and the number of cycles in a second is a mere matter of arithmetic. There is much greater difficulty in realizing the electrical and magnetic structure of light. It is not possible here to give more than the briefest statement. Maxwell perceived from the relations of electrical and magnetic actions that an electro-magnetic "wave" should travel at the speed of light in air or in a vacuum, a speed vastly higher than any that can be accounted for by mechanical properties of matter. He saw that these waves might have any wave-length and any corresponding number in a second, but that whatever size the wave might be it would always travel at the same speed, the speed of light. Within its octave light visible to the eyes of man has various wave-lengths, but travels at the same enormous speed. With suitable instruments, such as the spectroscope, the thermopile and the photographic plate, kinds of light, which are light in all respects except that the eye is not affected by them, may be traced for many octaves above and below the narrow limits perceived by the eye. All these come with visible light from the sun and other sources and at the same speed. Maxwell had no way of making an electro-magnetic wave, but he recognized it as an electrical necessity if only means existed to create it, and when made he knew that it would be of the same kind as the light wave. It remained for Hertz to discover how these waves could be produced by purely electrical means, and when made they were found to behave as predicted by Maxwell, to be in fact light waves but of much greater length.

Now, thanks to Marconi and others, these waves are well known, they are made for the purpose of telegraphing from ship to ship 300 or 600 metres long, and for long-distance telegraphy without wires two or three miles long and consequently at the rate of only 100,000, or even less per second, and in the laboratory the electrically produced waves can be made so short that they only fail to reach the longest waves in the whole spectrum by an octave or two. When we realize all this we are able to speak of the electro-magnetic structure of light with confidence.

Now returning to the soap-bubble and remembering that it is very thin it is not difficult to understand that light which has been reflected partly from its front surface and partly from its back surface may be so split up as to have one part which has travelled one or two or three hundred-thousandths of an inch further than the other; then whether light will be seen or not will simply depend upon whether the two parts are in electrical and magnetic agreement or the reverse.

It is necessary to understand that while the wavelength in air of the green light is $\frac{1}{50000}$ inch, in water where light only travels at three-quarters of the speed that it does in air it will only travel $\frac{3}{4}$ of $\frac{1}{50000}$ inch, that is $\frac{1}{66667}$ inch before the phase is reversed twice. The wave-length of any colour in water is three-quarters of the wave-length of the same colour in air.

With this introduction I hope that the examination of the action of a soap-film on light will be less difficult than it would be if all these considerations had to be introduced at once.

I have already said that the "wave-length," that is, the distance in which the polarity is reversed and reversed back to its former state again, in the case of a particular green light is $\frac{1}{50000}$ inch or 20 millionths of an inch. The wave-length of the yellow of the spectrum is longer and red longer still, while that of the blue is shorter and of the violet shorter still. All these and all the intermediate colours and wave-lengths combined together constitute the white light of day, and if any one or more are removed from white the remaining colours combined produce a colour which is called the complementary colour to the mixture of those that are removed. If

we can see how a soap-bubble may fail to reflect
certain colours and yet reflect others, we shall understand how the colour that we see reflected is the
complementary colour of those that it does not reflect.
Instead of considering white light it will be simpler
at first to take the case of some simple colour and so
avoid the complication that would result from considering all the colours and all their wave-lengths at
once.

Let us therefore take the green light so often referred
to, with a wave-length of $\frac{1}{50000}$ inch. Let me repeat
the polarity of this coloured light is reversed every
hundred-thousandth of an inch, so that if the light is
split into two beams following the same course and one
is delayed relatively to the other 1, 3, 5 or an odd
number of hundred-thousandths of an inch the combination of the two beams produce darkness; if the
delay is 2, 4, 6 or an even number then the two reinforce one another and the green colour is seen of
greatly increased intensity.

The matter will be clearer if we take an example.
Fig. 77 shows on a very greatly magnified scale a film
of water between the lines 1, 1 and 2, 2. A scale of
millionths of an inch is placed below. Consider a
beam of the green light between *ab* and *cd* travelling
in the direction shown by the arrow heads, striking
the first surface 1, 1 of the film between *b* and *d*.
Then in general it will split up into two beams, one
entering the film and being refracted to *ef*, and the
other reflected at *bd* to *kl*. The proportion of the
light which is found in the two beams depends upon
the inclination of the beam and also on its polarization, if any, which is an expression defining the
directions in which the electrical and magnetic actions
occur. For a steep inclination such as is shown in
the figure the reflected beam is much less intense
than the refracted beam, but as the light strikes

the film more and more obliquely a greater proportion of light in general is reflected. There are complications depending on the state of polarization, if any, of the light, but into these it is not necessary to enter. The beam which has struck the second surface 2, 2 at *ef* is there again split up into two beams, one passing out into the air again and being refracted to *gh* and one

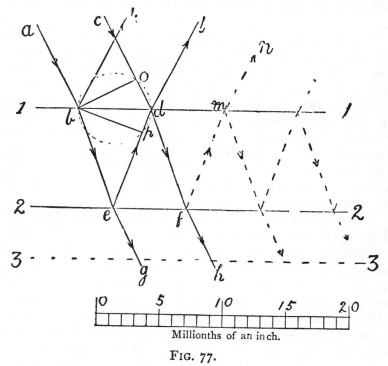

FIG. 77.

reflected to *dm*. Here again the process of splitting is repeated. Some passes on into the air to *ln* while some is reflected back into the film, and so the process is repeated indefinitely. As the light which is reflected from the first surface to *kl* is far greater than all the rest of the light which emerges from this surface I shall consider it alone, but strictly all the emergent light paying due regard to its phase and intensity must be taken into account to make the theory complete.

Following then a construction analogous to that of Fig. 76, and drawing the dotted circle about bd as a diameter, and joining b with o and p, the points where the dotted circle cuts the lines cd and ed, we have do and dp distances in air and in water which light takes the same time to traverse; we have bo at right angles to ab and cd, and as this line is all at the same distance from the source of light, the light at all parts of this line is in the same phase. Now the two rays from a and c are in the same phase at b and o, thereafter some of the a ray goes to e, and thence via p and d to l. Some of the c ray after leaving o goes via d to l. As the times of travelling from o to d and from p to d are equal, and thereafter the two follow the same course up to l, the delay introduced into the a ray is the time of travelling from b via e to p. Now the thickness of the film between the lines 1, 1 and 2, 2 and the inclination of the original beam of light to the water-film are so chosen that this distance bep is exactly one wave-length *in water*, in the very highly magnified drawing. It would be expected then that the polarity would be reversed and reversed again back to its original state in this distance, and so as od and pd are equivalent paths, the two rays would arrive at d and travel up the common path to l in the same phase. If so they should reinforce one another, and the film should appear brilliantly green to an eye placed to receive the reflected light. It should appear green, that is, if the light thrown on to the film is green.

There is, however, a difference in the conditions of the reflection at the first and second surfaces of the water-film. At the first surface light is reflected from the material in which it travels more slowly, while at the second it is reflected from the material in which it travels more quickly. Now it is a general property of wave reflection, whether mechanical or electrical in its mechanism, that when the two rays are reflected under these different conditions there is a reversal of phase of one ray with respect to the other as compared with that which the mere consideration of distance travelled would indicate. The result is that in Fig. 77, the two rays which each would traverse *dl*, if the other did not exist, are actually in opposite phases, and no green light can be seen. The two rays may be considered samples of corresponding pairs all over the film if it is of the same thickness, and so the whole of the light that would be reflected from the first surface, if there were no light from the second surface, is in the opposite phase to the whole of the light reflected from the second surface, and that would come out through the front surface if it alone existed. As these actions are exactly equal and opposite no light is seen reflected from the film. What then becomes of the energy contained in these two beams of light? Each of them contains as measured in foot-pounds per minute or in horse-power, it is true by a very small number, a certain definite amount of energy. This cannot be lost. When waves "interfere," as this process is called, and produce darkness or silence or a state of rest, the energy which has disappeared from the place where action seems to be destroyed, reappears of necessity somewhere. Now referring again to Fig. 77, I have stated that as between the first and second surfaces there is this difference, that the two rays, either of which would travel along *dl* if it were not for the other, are relatively to one another reversed in phase. It does not matter which is reversed, or whether each is altered a little so as to make them opposite in phase ; along *dl* they are opposite, and so there is no reflected ray *dl*. Again it does not matter how the reversal is effected, the ray *df* as the result of double reflection within the film will be in the same phase as the other ray *df* produced by *cd* entering the film. These two being in the same phase join and become an extra strong ray, so a more intense green light will pass out along *fh*, and the film is more transparent to green light.

If the film had been half as thick again, so that the second surface was in the position shown by the dotted line 3, 3, then there would have been a loss of three half waves within the film, and this combined with the phase reversal would leave the corresponding rays *dl*, *dl* in the same phase, and green light would be brilliantly reflected. At the same time the twice reflected ray *df* would be opposite in phase to the stronger refracted ray *df* coming from *c* direct, and so the ray *dfh* from *c* would be weakened by the addition of the twice reflected ray, and the film would seem less transparent.

Similarly if the film had been twice, three times or any whole number of times as thick as the film shown in Fig. 77, green light would not be seen reflected, whereas if it were half as thick, one and a half or any-whole-number-plus-a-half times as thick, then green light would be brilliantly seen, that is of course if as is so far supposed green light only is falling on the film. For intermediate thicknesses there is only partial interference, and the reflected light is of intermediate intensity.

It is interesting to consider what must happen if the film is so thin that the wave length is very long by comparison with the thickness, say forty or fifty times as great; then the thickness may be considered very nearly equal to the whole number 0, and it should agree with the rule of the whole numbers just given, and fail to reflect green light. This actually is the case, and soap-bubbles are easily prepared so thin as to appear black by comparison with the rest of the film, unless the whole is black, and in that case in a good light the film can just be seen. Now suppose instead of green light we take light of a different colour and wave-length, *e.g.* a particular red with a wave-length of $\frac{1}{37037}$ or twenty-seven millionths of an inch in air, then if the film of Fig. 77 is supposed to be thickened in the same proportion as that by which the wave-length of red is greater than the wave-length of green, that is in the proportion of 27 to 20, whenever green light would be reflected or would fail to be reflected with the thinner film, red light would be reflected or would fail to be reflected with the thicker film. Similarly if blue light be taken, which has a shorter wave-length than green, a correspondingly thinner film will give the same result again. And so for every colour of the spectrum.

The total range of wave-lengths in the visible spectrum is nearly an octave, that is the extreme red has a wave-length nearly double that of the extreme violet, and the green is about in the middle.

The following figures are examples—

	Wave-length in air in millionths of an inch.	Number in an inch in air.	Wave-length in water in millionths of an inch.
Limit of Red ...	30	33,333	22·5
Red	27	37,037	20·3
Yellow ...	23	43,478	17·2
Green	20	50.000	15·0
Blue	18	55.555	13·5
Violet	17	58,824	12·8
Limit of Violet	15½	64,516	11·6

From what has been said then it will be clear that when white light falls upon a soap-film the consideration of what will be seen is not simple, for every colour and wave-length is acted upon differently. The simplest case is that of the very thin film which cannot reflect any of the colours, it is simply black. We shall see how to prepare these later. Suppose the film gradually to increase in thickness, then all the colours begin to increase in brilliancy, more especially the violet and blue, and the light that is reflected has at first a tinge of blue. As it gets thicker the blue and violet which have the shortest wave-lengths are the first to be cut out, that is, when the thickness is about 6 millionths of an inch.   Now the red is being reflected most strongly and the yellow and the green less strongly, and the result is the well-known straw colour of tempered steel.   As the film gets thicker the green light fades away, and the straw colour darkens to brown, and these change to violet and blue as the yellow and red in turn are wiped out.   As the film gets thicker still each of the colours weaken, vanish and grow again in turn, but all at different rates, and it becomes very confusing to follow the process.   This is greatly simplified by the use of the coloured plate,* which is so constructed that you can see at a glance what colours are present in the reflected light from a soap-bubble of any thickness from nothing up to 50 millionths of an inch, that is, if the light strikes and is reflected from the bubble perpendicularly or nearly so.   The horizontal shades of colour represent the colours of the spectrum.   The black lines lettered *B, C, D, E, b, F, G* are the more conspicuous dark lines in the solar spectrum with their usual designations, and these serve to define the colours by reference to a real spectrum of daylight.   At the left-hand end are scales of wave-lengths in air of the range of the visible spectrum, and at the right-hand end are the corresponding wave lengths in water.   The scale below is a scale showing the thickness of the bubble in millionths of an inch, and the shaded bands, which become more and more oblique towards the right, show, for any thickness of film, which colours are absent or are present in their full brilliancy or are partly obscured in the reflected light.   Taking a bubble of thickness 4 millionths of an inch and holding a straight-edge vertically across the diagram from the division of the lower horizontal scale corresponding to the thickness 4 it will be seen that it lies wholly in a bright band, showing that all the colours are nearly equally present.   This then is the white of

---

* Reproduced on the back cover.

the first order. As the edge is slowly advanced to the right it will be seen how the shaded area in the level of the violet and blue begin to be crossed. The sum of all the remaining colours, due allowance being made for the intensity of each as represented by the depth of shade, gives the colour that is seen in the soap-bubble, and this is the straw of tempered steel represented in the colour band above the spectrum. By the time the straight-edge has reached the division corresponding to the thickness 10 millionths of an inch the violet and blue are of their full intensity, the green is weak, and yellow and red are gone. This is the blue of tempered steel ; the same colours are seen in the bubble. As the straight-edge moves further on to the right, two, three, and more bands are successively intersected, and the resultant colour becomes more and more difficult to ascertain. Actually the colour of a bubble of variable thickness may be examined with a spectroscope, and then the spectrum is seen intersected by one or more inclined bands, and the shaded part of the diagram or some of it is seen with all the colours true and brilliant. No imitation with pigments can compare with this. The film colours of the first few orders that are near the left-hand end of the diagram and which are imitated crudely in the upper band of colour are the most brilliant. When the colour is made up of too many patches distributed all down the spectrum the colours get paler and degenerate into pale shades of red and green. I would only remark here that you cannot judge of the colour impression produced by a mixture of colours by mixing pigments out of a box of paints. Thus blue and yellow mixed as pigments produce green, while blue and yellow added together produce the impression of white. The colours of pigments are really the result of subtracting the complementary colour from white. By taking two pigments and mixing them more is subtracted from white and only that which both contain is seen. Blue and yellow pigments both contain green, hence green is seen when the pigments are mixed.   Blue and yellow added together, which may be done by painting blue and yellow sectors on a card and spinning it, using, however, much more yellow than blue, appear white, as either between them they contain all the colours that go to make up white, or pure blue and yellow alone may produce the sensation of white.

The series of colours and corresponding thicknesses illustrated in the coloured plate, are calculated on the supposition that the light falls on the film and is reflected from it vertically or very nearly so, and they agree in position with the scale of colour as determined by Reinold and Rücker.   If, as usually happens, the light falls on the film obliquely that which is reflected from the second surface will traverse the film twice obliquely instead of twice straight across.   It will therefore travel a greater distance in the film, and the natural expectation is that it would be retarded more and the film would show the colour of a thicker film seen vertically.   This however is not the case.   With increased obliquity, while the length of the path *bed* (Fig. 77) within the film is certainly increased it must be remembered that the piece *pd* within the film is not counted, as it balances the piece *od* in the air, and the distance *bep* only has to be considered as the measure of the delay.   Now with increased obliquity the distance *od* rapidly increases, the corresponding distance *pd* similarly increases, in fact it increases faster than the total distance *bed* within the film increases, and so the remaining length *bep*, which is all that matters, actually gets rather less.

To what extent this is the case can be seen very clearly from the construction shown in Fig. 78, in which the film 1, 1 2, 2 of Fig. 77 is drawn in thick lines together with the lines *be, ed* and *bp*.   Now suppose 2, 2 to be a mirror (as it is) and that all above it is seen reflected in it below as shown by the thin lines.   Then by very elementary geometry *bep′* is a straight line and *b′p′* is at right angles to it.   Also the distance *bep*, which we wish to know for all possible inclinations of *be* within the film, is equal to the straight line *bep'*, so if we can see how *bep'* varies in length as its inclination is varied we shall know all we want. As the angle *bp'b'* is a right angle, a semicircle drawn on *bb'* as a diameter will pass through it. Draw the semicircle therefore, and then whatever inclination *be* may have the length of the line *bep'* cut off by the semi

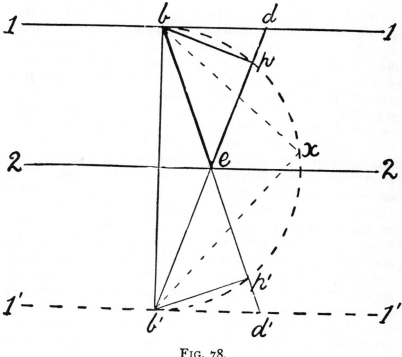

FIG. 78.

circle is a measure of the delay between the two inter-fering rays. It will be noticed that as the inclination is increased the chord *bp'* will at first diminish very slowly indeed, and so the colours of a soap-film change very little when seen at or near the vertical. As the in-clination of the line *bp'* to the film becomes less its length diminishes more and more rapidly, and so with oblique illumination the colours of a soap-film not only are those that a thinner film would show with vertical illumination but the change becomes increasingly notice-able ·as the obliquity becomes greater. As, however, the inclination of bp' changes very little with consider-able change of the inclination of the ray in the air where this is very oblique, it comes about that the quickest change of colour occurs when the external ray makes an angle of $39°$ with the film. As, accord-ing to the construction of Fig. 77, dp cannot be more than $\frac{3}{4}$ bd, so by similar triangles $b'p'$ cannot be more than $\frac{3}{4}$ bb', and so $b'x$ equal to $\frac{3}{4}$ bb' is the greatest possible length that $b'p'$ can attain, and bx making an angle of about $41°$ with the film is the least possible inclined ray that can enter the film, and its length, which may be shown to be $\dfrac{\sqrt{7}}{4}$ of bb' or very nearly $\frac{2}{3}$ bb', is a measure of the smallest possible delay for a ray which has entered the film. Thus it is that with a grazing incidence of the light outside a soap-film cor-responding to the ray bx within the film the colour is that which would be given by a film of two-thirds the thickness seen vertically. The following table shows the factor by which the thickness of a film seen obliquely must be reduced in order to give the thickness that would give the same delay with vertical illumination. By the use of this table and the coloured plate the thick-ness of any soap-bubble if the colour can be recognized may be determined. As the thickness determines the weight I have added to the plate at the top a scale show-ing the weight of a square inch of a film of any colour in thousandths of a grain. The same scale gives the weight of a square centimetre in thousandths of a centigramme or hundredths of a milligramme and the thickness in ten thousandths of a millimetre.

Table showing the fraction by which the thickness of a soap-film seen obliquely is to be multiplied in order to give the thickness of a film that would give the same colour with vertical illumination :—

Inclination of ray in air.	Inclination of ray in the film.	Fraction.
0°	41° 25'	·6615
10°	42° 23'	·6741
20°	45° 11'	·7094
30°	49° 30'	·7604
40°	54° 56'	·8185
50°	61° 11'	·8762
60°	67° 59'	·9271
70°	75' 8'	·9665
80°	82° 31'	·9915
90°	90° 0'	1·0000

As three things may change, and when different parts of a spherical bubble are looked at do all change at once, namely the wave-length, the thickness and the inclination, it is not surprising that there is so great a wealth and variety of colour and that the out-of-door bubbles already described are so attractive. For the same reason a spherical bubble is not the most convenient to employ when comparing the actual fact with the result of the theory as I have given it.

For this purpose a plane film on a metal ring is suitable, and it is most convenient to make the ring with three feet as a tripod so that it will stand level, or nearly so, and a glass shade should be provided to put over it to keep off draughts and diminish evaporation. If there is glycerine in the liquid the thinning process is much more slow than it is with a pure soap-and-water solution. With this last it is best to put wet black paper on a plate below the film or in the shade, and the background against which the film is viewed should be black or as dark as possible, and if this is shaded from the light as well so much the better.   It is best to make observations in a window facing away from the sun, and a sky with white clouds is best.

In order to stretch a film over the ring dip it in soap solution and place it so that it is not quite level and cover it with the glass shade immediately.   Instead of dipping stroke a piece of thin celluloid which has been wetted with the soap solution over the ring.   As the film gets thinner owing to the process of draining, the higher-order colours, pale pinks and greens, will appear across the film, and at the highest level, the lower-order colours of brighter hues will be seen successively.   The paler tints pass gradually down the film and the brighter bands widen out and the succession of colours imperfectly set out at the top of the coloured plate will gradually widen out and pervade the film. If there is no glycerine in the soap solution black spots will very soon appear in the white, and very often the formation of one is followed by a host of others in rapid succession, and a larger area of black is formed. When black, the film is either $\frac{1}{2000000}$ or $\frac{1}{4000000}$ inch thick and is very tender.   The thinner film is the blacker of the two.   These thicknesses obviously cannot be determined by mechanical means.   Reinold and Rücker, Stansfield and Drude have all made accurate observations on these films.   Using two optical and one electrical method these observers have between them found the thickness to be as just stated.   One of the optical methods depends on interference of two beams of light, one or other of which passes through a great number of black films which optically lengthen the path by a third of their combined thickness.   Drude has kept his black films under observation as long as three weeks, but unless the air round them is as moist as the soap solution can make it the black film breaks very soon in consequence of evaporation.   Going back to the coloured part of the film the order in which the colours appear is so to speak backwards, those due to the thicker films appearing first. After several very pale pink and bluish or greenish bands have appeared a salmon pink and bluish green pair of bands seem quite conspicuous and highly coloured. The film is then if seen at a steep angle from 25 to 30 millionths of an inch thick. The bluish green is succeeded by a reddish purple, and this at once distinguishes it from the thinner and brighter apple green which is followed by blue and then by reddish purple. The yellow that follows this is the only good yellow in the whole series of colours. This changes to blue through an intermediate and imperfect white when the film is about 12 millionths of an inch thick. This blue and the succeeding purple, brown and straw colour are the colours of tempered steel formed in the same way in the thin film of oxide. While this blue and purple in the soap-film vie with the convolvulus in the glory of its colour, it is curious that when the bands of colour are very narrow this tempered steel group appears like a black band between the two white bands or like a black band with a fiery copper-coloured edge on the thin side. The colours of the thicker parts of the soap-bubble are pale or barely visible, not because there is any failure in the operation of the interference as described, but because, as the coloured plate makes clear, the higher-order colours are the result of combining so many patches of colour evenly distributed along the spectrum. It is easy to show that the interference goes on to far greater thicknesses than 50 millionths of an inch, at which the diagram comes to an end, by using instead of white light, light of one colour only. This is imperfectly but most easily effected by looking through a sheet of ruby glass, such as is used by photographers. Immediately the pale pinks and greens become conspicuous reds and blacks, and these bands are continued into parts of the film which show no colour when observed directly. A far better source of light is the yellow flame obtained by holding a salted wick so as just to touch the base of the flame of a spirit lamp or of a bunsen gas burner. With this light nearly 500 bands may be seen. The light which passes through red glass covers a certain range in the spectrum even though only red light passes through it, the orange red has a shorter wave-length than the deep red. When therefore the soap-film is perhaps ten times as thick as that which produces the first dark band in the red, say as much as $\frac{1}{10000}$ inch, the eleventh bright band of the shorter wave and the tenth dark band of the longer wave may be formed together and the bands cease to be visible. In the same way when the film is about 500 times as thick as that which gives the first yellow dark band, the two colours of the salt vapour which differ in wave-length by about 1 in 1000 are in opposite phases and the bands disappear. With a film of double this thickness, however, they are in the same phase again and the bands re-appear. Even though the bands disappear, that is are not visible to the eye because the two yellow colours in the salt flame cannot in this way be distinguished by the eye, they are there all the same and with sufficiently delicate spectroscopic means they could still be made visible, but I am not aware that any one has done this.

The colours of thin films which have been described with reference more especially to a water-film, are often seen in thin films other than water. The motor-car has brought oil on to the roads in all our towns, and when these are wet large areas of highly coloured oil-film may often be seen. Cracks in glass often show these colours, and pressure will cause the colours to change as the crack opens and shuts a few millionths of an inch. It is pretty obvious that it is not possible to measure the thickness of a soap-film with any kind of screw micrometer or calliper, but the fact that the colours and thicknesses of air- or water-films between glass surfaces agree as indicated by the theory of interference may be proved by observing them in the thin film between a very slightly curved lens, and a glass plate when the thickness at any known distance from the point of contact may easily be calculated. These are the rings observed by Sir Isaac Newton. It may be worth while to mention here that the part of the theory which at first appears unconvincing, namely the phase reversal resulting from the two kinds of reflection, causing the dark and bright bands to change places as from the positions which they would occupy if delay only operated, may be confirmed experimentally.

By choosing two solids through which light travels, one as slowly and the other as quickly as possible, and a liquid through which light travels at an intermediate speed, and making a thin film of the liquid between the two solids there is no change of condition in the two reflections. Light either travels progressively more slowly or more quickly in passing through the three materials. If then one or other condition of reflection gives rise to an abrupt reversal of phase, then this reversal will occur either not at all or twice according to the direction of the light through the three materials, and either way there will be no ultimate reversal, and the centre spot of a Newton ring system will be white and not black as it is when a film of air is between two glass surfaces. The material that is commonly met with through which light travels most slowly is diamond. Light takes about two and a half times as long to pass through a piece of diamond as it does to travel the same distance in a vacuum. Next to diamond is the flashing gem stone zircon or jargoon, the time of which is 1·9. Then follow the ruby and sapphire, and after that ordinary kinds of glass. The flat table of a brilliant cut diamond or zircon may therefore be used for the lower surface. Of liquids, the evil-smelling and highly inflammable carbon bisulphide is the best, as light takes 1·63 times as long to pass through this as through a vacuum, but oil of bitter almonds or aniline with times of 1·6 and 1·57 respectively both answer perfectly. Then for the upper material a common glass spectacle lens convex and of little power with a time 1·5 does well, and it is easily obtained. To make the experiment fix a ring or bracelet with a brilliant cut stone upwards. On the upper face place a minute drop of the liquid and then press one face of the lens near its edge on to the face of the gem, thus squeezing the liquid into a thin film. The reason for using a part of the lens near the edge is that there the upper face of the lens is not quite parallel to the lower face, and light from a restricted source may be reflected from the face of the gem into the eye while light reflected from the upper surface of the lens misses the eye. Then if the place of contact is clean, and if it is examined with a good pocket-lens a small ring system will be seen with a white centre. With the same appliances but without the liquid the ordinary Newton rings with a black centre will be seen of a far greater brilliancy. If instead of aniline water is used as an intermediate fluid the centre of the ring system will be seen black as it should be. There is a curious appearance which may be seen with aniline or with water, but not with carbon bisulphide. While pressing the lens upon the gem with the fingers the point of contact moves about, and when this is the case a brilliant white spot follows the centre spot. This is caused by a small bubble of vacuum, for the liquid cannot follow the movement, slow as it actually is, and insinuate itself quickly enough into a crack less than $\frac{1}{100000}$ inch thick. Carbon bisulphide is so mobile that this liquid follows perfectly. The other extreme is glycerine, which at each movement leaves vacuum bubbles large enough to enclose several rings, and then these can be seen on the one side and the smaller corresponding rings in the glycerine can be seen on the other side of the central spot.

While in discussing the theory I merely stated that as

a fact there was a reversal of phase relatively one to the other in the rays reflected from the air-water and the water-air surfaces and that it did not matter whether the reversal took place all at one surface or all at the second, or partly at one and partly at the other, the result was nevertheless such that the two rays always showed relatively to one another a reversal of phase as compared with what they would show if the extra path of one alone was operative. The late Sir George Stokes showed by a process of pure reasoning that this must be so. This part of his interesting theorem requires no mathematical expression, and I can therefore give it here. The theorem is based on the principle that if a beam of light which has been broken up into two or any number by reflection, refraction or other action not involving absorption could be reversed in direction so as to retrace its path, the original beam and no other would be exactly reproduced.

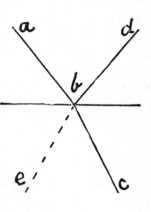

FIG. 79.

The beam *ab* Fig. 79 on striking water obliquely is divided into two, *bc* weaker than *ab* refracted into the water and *bd* weaker than *ab* reflected back into the air. Now if these two are reversed they must reproduce *ab* reversed or *ba* and no other. *db* alone, however, would make a beam *ba* weaker than itself and a beam *be* weaker than itself. *cb* would do likewise. As *be* does not exist the contributions which each make to *be* must be equal and opposite, leaving their full values free to go into *ba* and reproduce it. As each beam in the supposed operation of contributing to *be* enters water once that part of the operation is the same for both. On the other hand each beam is reflected once, but one reflection is from water while the other is from air, and since the complete result is absence of *be* the phases resulting from the two kinds of reflection must be opposite and the intensities equal. Further, since the two contributions to *ba* reproduce *ab* of its original intensity they must be in the same phase. As one of the contributing beams has been reflected twice while the other has made an entry into and an exit from water, these two pairs of operations must have produced either no change of phase or an equal change of phase, and the intensities must be such that the energy travelling up *ba* after two reflections from water added to that which is left after one entry and one exit will amount to the energy in the original beam. When in this argument entries and exits and the two kinds of reflection are spoken of the relations are only true for such pairs of inclinations to the film in air and in water as result from refraction, not for any inclinations nor even for equal inclinations. The applications of these results to the colours of Newton's rings is not altogether simple, because, as is explained on p. 145, the complete theory of these rings requires that account should be taken of all the light which emerges from within the film and not only of the first beam reflected from the second surface. It is only then that the first and second surface reflections are exactly equal to one another in amount. The first of the reflections from the second surface has, however, the preponderating influence, and Stokes's theorem just given at least shows the reversal of phase as between the two classes of reflection, and so it is that the central spot is dark and not light in the usual case. I would only add that in evaluating the intensity after a succession of identical operations this is reduced in the same proportion, while the phase is changed by the same amount, each time.

The actual process of draining of the soap solution between the two surfaces of a film is excessively slow, and with thin highly coloured films slow beyond all expectation. Prof. Willard Gibbs has calculated the rate at which the liquid will drain in vertical films of very varied thickness.

Thickness of Film.	Rate of Downward Movement.
400 millionths of an inch.	$\frac{1}{250}$ inch a second.
40 ,, ,,	$\frac{1}{150}$,, a miuute.
4 ,, ,,	$\frac{1}{50}$,, an hour.

If then there were no evaporation and no other cause of thinning, the process of reaching the black stage would be far greater than it is.

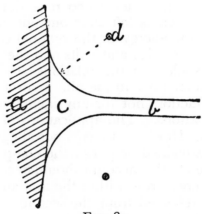

FIG. 80.

Perhaps the most interesting of the methods by which the plane film held in a ring gets thinner and one which when understood is extremely interesting to watch, especially with a good lens, was described by Prof. Willard Gibbs. Suppose *a*, Fig. 80, to be a section of the side of the wire and *b* to be a section of the film on a very highly magnified scale, then where the edges of the film meet the wire they do not meet it abruptly or at right angles but bend round as shown in the figure to meet it tangentially. Now the size of these curves simply depends on the amount of liquid which will remain at all in the first instance. When the curved triangle c bounded by the wire and the two water surfaces is large the liquid within it drains from the wire very quickly, and it soon assumes a size which the eye shows to exist but to be excessively small. Now suppose the radius of curvature shown dotted from d to be $\frac{1}{1000}$ inch, and it need not be more, the curved stretched film will exert a suction on the liquid in the triangular space. The extent of this suction is easily ascertained as it is the same as that at the curved surface of water lifted by capillarity (see p. 26) between two parallel plates twice this distance or $\frac{1}{500}$ inch apart. The weight of water lifted in a film one inch wide between two such plates is equal to the upward pull due to capillarity on the plates, and this in the case of soapy water is just over $1\frac{1}{4}$ grains to the linear inch on each plate or $2\frac{1}{2}$ grains in all. What length of water trough one inch wide and $\frac{1}{500}$ inch deep will contain $2\frac{1}{2}$ grains weight of water? A cubic inch of water weighs $252\frac{1}{4}$ grains. One five-hundreth of this is almost exactly half a grain and therefore five inches of the trough will be needed to contain $2\frac{1}{2}$ grains. The water therefore would rise to a height of five inches between the plates, and the suction is therefore that due to a drain pipe filled with water dipping into water five inches below the triangle. This is the suction within the triangle, but at b where the films are plane, though equally tightly stretched, they exert no suction. The triangular thickening therefore of the film along the wire acts as a constant suction to the film. If the film is set exactly level the operation of this suction can be readily observed under a lens. Small highly-coloured pear-shaped patterns develop at very regular intervals all round the film. If the film is slightly inclined these coloured and lighter places, lighter because thinner, tend upwards and the continuous formation of coloured tadpole-shaped figures which swim to the top one after the other is one cause of the formation of the thin areas at the upper part of the film. The colours of these figures are always those of a thinner film than that in which they grow, and when this film is very thin even black figures with coloured edges may be seen.   It must be remembered that just as when liquids are dealt with in bulk the lighter as measured say in pounds to the gallon or grains to the cubic inch float on the heavier, *e. g.* oil floats on water and both float on mercury, so in these films, where closely adjacent parts show no disposition to mix and average their thickness and where all the liquid has the same density as measured in grains to the cubic inch, the weight of a square inch of one coloured film is different to that of a square inch of another coloured film ; the thinner and therefore lighter parts tend to float through and above the thicker and heavier parts and so sort themselves in layers of the same thickness and colour.   It is in scientific language a case of surface density instead of volume density.   I have sought to prevent the formation of the coloured edge patterns by making a ring with an edge so fine that I confidently expected it would, so to speak, split a bubble. I have had the advantage of seeing the whole process of manufacture of the blades of the Gillette razor, and I know of no edge which is equal to this in the perfection of its thinness.   On making up a square ring of four new Gillette blades just meeting at the corners I found that the formation of the coloured patterns in the film along the edge was very much reduced as compared with that along a wire or a roughly turned edge, but I could not entirely get rid of them.

The examination of a thinning and slightly inclined film is so fascinating and the variety of patterns and movements is so great that it is impossible to describe all that may be seen.   I would only urge any one who has taken any interest to set up the very simple apparatus required, and whether using soap and water, soap and water containing some glycerine, or the carefully pre- pared and pure mixture described on p. 170, there will be abundant reward.

PRACTICAL HINTS

I HOPE that the following practical hints may be found useful by those who wish themselves to perform successfully the experiments already described.

Drop with India-rubber Surface

A sheet of thin india-rubber, about the thickness of that used in air-balls, as it appears *before* they have been blown out, must be stretched over a ring of wood or metal eighteen inches in diameter, and securely wired round the edge. The wire will hold the india-rubber better if the edge is grooved. This does not succeed if tried on a smaller scale. This experiment was shown by Sir W. Thomson at the Royal Institution.

Jumping Frame

This is easily made by taking a light glass globe about two inches in diameter, such, for instance, as a silvered ball used to ornament a Christmas-tree or the bulb of a pipette, which is what I used. Pass through the open necks of the bulb a piece of wire about one-twentieth of an inch in diameter, and fix it permanently and water-tight upon the wire by working into the necks melted sealing-wax. An inch or two above the globe, fasten a flat frame of thin wire by soldering, or if this is too difficult, by tying and sealing-wax. A lump of lead must then be fastened or hung on to the lower end, and gradually scraped away until the wire frame will just be unable to force its way through the surface of the water. None of the dimensions or materials mentioned are of importance.

Paraffined Sieve

Obtain a piece of copper wire gauze with about twenty wires to the inch, and cut out from it a round piece about eight inches in diameter. Lay it on a round block, of such a size that it projects about one inch all round. Then gently go round and round with the hands pressing the edge down and keeping it flat above, until the sides are evenly turned down all round. This is quite easy, because the wires can allow of the kind of distortion necessary. Then wind round the turned-up edge a few turns of thick wire to make the sides stiff. This ought to be soldered in position, but probably careful wiring will be good enough.

Melt some paraffin wax or one or two paraffin candles of the best quality in a clean flat dish, not over the fire, which would be dangerous, but on a hot plate. When melted and clear like water, dip the sieve in, and when all is hot quickly take it out and knock it once or twice on the table to shake the paraffin out of the holes. Leave upside down until cold, and then be careful not to scratch or rub off the paraffin. This had best be done in a place where a mess is of no consequence.

There is no difficulty in filling it or in setting it to float upon water.

Narrow Tubes and Capillarity

Get some quill-glass tube from a chemist, this is, tube about the size of a pen. If it is more than, say, a foot long, cut off a piece by first making a firm scratch in one place with a three-cornered file, when it will break at the place easily. To make very narrow tube from this, hold it near the ends in the two hands very lightly, so that the middle part is high up in the brightest part of an ordinary bright and flat gas flame. Keep it turning until at last it becomes so soft that it is difficult to hold it straight. It can then be bent into any shape, but if it is wanted to be drawn out it must be held still longer until the black smoke upon it begins to crack and peel up. Then quickly take it out of the flame, and pull the two ends apart, when a long narrow tube will be formed between. This can be made finer or coarser by regulating the heat and the manner in which it is pulled out. No directions will tell any one so much as a very little practice. For drawing out tubes the flame of a Bunsen burner or of a blow-pipe is more convenient; but for bending tubes nothing is so good as the flat gas flame. Do not clean off smoke till the tubes are cold, and do not hurry their cooling by wetting or blowing upon them. In the country where gas is not to be had, the flame of a large spirit-lamp can be made to do, but it is not so good as a gas-flame. The narrower these tubes are, the higher will clean water be observed to rise in them. To colour the water, paints from a colour-box must not be used. They are not liquid, and will clog the very fine tubes. Some dye that will quite dissolve (as sugar does) must be used An aniline dye, called soluble blue, does very well. A little vinegar added may make the colour last better.

Capillarity between Plates

Two plates of flat glass, say three to five inches square, are required. Provided they are quite clean and well wetted there is no difficulty. A little soap and hot water will probably be sufficient to clean them.

Tears of Wine

These are best seen at dessert in a glass about half filled with port. A mixture of from two to three parts of water, and one part of spirits of wine containing a very little rosaniline (a red aniline dye), to give it a nice colour, may be used, if port is not available. A piece of the dye about as large as a mustard-seed will be enough for a large wine-glass. The sides of the glass should be wetted with the wine.

Cat-Boxes

Every school-boy knows how to make these. They are not the boxes made by cutting slits in paper. They are simply made by folding, and are then blown out like the "frog," which is also made of folded paper.

Liquid Beads

Instead of melting gold, water rolled on to a table thickly dusted with lycopodium, or other fine dust, or quicksilver rolled or thrown upon a smooth table, will show the difference in the shape of large and small beads perfectly. A magnifying-glass will make the difference more evident. In using quicksilver, be careful that none of it falls on gold or silver coins, or jewellery, or plate, or on the ornamental gilding on book-covers. It will do serious damage.

Plateau's Experiment

To perform this with very great perfection requires much care and trouble. It is easy to succeed up to a certain point. Pour into a clean bottle about a table-spoonful of salad-oil, and pour upon it a mixture of nine parts by volume spirits of wine (not methylated spirits), and seven parts of water. Shake up and leave for a day if necessary, when it will be found that the oil has settled together by itself. Fill a tumbler with the same mixture of spirit and water, and then with a fine glass pipe, dipping about half-way down, slowly introduce a very little water. This will make the liquid below a little heavier. Dip into the oil a pipe and take out a little by closing the upper end with the finger, and carefully drop this into the tumbler. If it goes to the bottom, a little more water is required in the lower half of the tumbler. If by chance it will not sink at all, a little more spirit is wanted in the upper half. At last the oil will just float in the middle of the mixture. More can then be added, taking care to prevent it from touching the sides. If the liquid below is ever so little heavier, and the liquid above ever so little lighter than oil, the drop of oil perhaps as large as a halfpenny will be almost perfectly round. It will not appear round if seen through the glass, because the glass magnifies it sideways, but not up and down, as may be seen by holding a coin in the liquid just above it. To see the drop in its true shape the vessel must either be a globe, or one side must be made of flat glass. An inverted clock shade with flat sides does well.

Spinning the oil so as to throw off a ring is not material, but if the reader can contrive to fix a disc about the size of a threepenny-piece upon a straight wire, and spin it round without shaking it, then he will see the ring break off, and either return if the rotation is quickly stopped, or else break up into three or four perfect little balls. The disc should be wetted with oil before being dipped into the mixture of spirit and water.

Other Liquids

I do not recommend the bisulphide of carbon mixture described in the text, owing to the smell and danger of using this material. Actually the orthotoluidine used by Mr. Darling is by far the most convenient, and it answers every purpose except that of breaking out into a ring when rotated. For this to succeed the more viscous oil originally used by Plateau is better than anything else.

A Good Mixture for Soap-bubbles

Common yellow soap is far better than most of the fancy soaps. Castille soap used to be very good, and this may be obtained from any chemist ; but olive oil from which it should be made is now generally mixed with cotton seed oil and this is not so good.

Bubbles blown with soap and water alone do not last long enough for many of the experiments described, though they may sometimes be made to succeed. Plateau added glycerine, which greatly improves the lasting quality. The glycerine should be pure ; common glycerine is not good, but Price's answers perfectly. The water should be pure distilled water, but if this is not available, clean rain-water will do. Do not choose the first that runs from a roof after a spell of dry weather, but wait till it has rained for some time, the water that then runs off is very good, especially if the roof is blue slate or glass. If fresh rain-water is not to be had, water that has been boiled and allowed to cool should be employed. Instead of Castille soap, Plateau found that a pure soap prepared from olive-oil is still better. When all the fatty acids except oleic acid are removed the resultant soap is oleate of soda. I have always used a modification of Plateau's formula, which Professors Reinold and Rücker found to answer so well. They used less glycerine than Plateau. It is best made as follows. Fill a clean stoppered bottle three-quarters full of water. Add one-fortieth part of its weight of oleate of soda, which will probably float on the water. Leave it for a day, when the oleate of soda will be dissolved. Nearly fill up the bottle with Price's glycerine and shake well, or pour it into another clean bottle and back again several times. Leave the bottle, stoppered of course, for about a week in a dark place. Then with a syphon, that is, a bent glass tube which will reach to the bottom inside and still further outside, draw off the clear liquid from the scum which will have collected at the top. Add one or two drops of strong liquid ammonia to every pint of the liquid. Then carefully keep it in a stoppered bottle in a dark place. Do not get out this stock bottle every time a bubble is to be blown, but have a small working bottle. Never put any back into the stock. In making the liquid *do not warm or filter it*. Either will spoil it. Never leave the stoppers out of the bottles or allow the liquid to be exposed to the air more than is necessary. This liquid is still perfectly good after ten years' keeping. I have given these directions very fully, not because I feel sure that all the details are essential, but because it exactly describes the way I happen to make it, and because I have never found any other solution so good.

In the year 1890 I had no difficulty in obtaining fairly good oleate of soda in this country. Now for several years I have tried what is called oleate of soda from one place after another and not a single specimen have I found that is the slightest use for the more delicate soap-bubble experiments. However on obtaining a quantity made by the well-known firm of Kahlbaum of Berlin, for whom Messrs. Griffin Bros. of Kingsway are agents, I found the oleate as tested by its bubble-blowing quality to be as good as or better than any I had met with before. If any one wishes to make his own oleate of soda then the following instructions kindly sent to me by Mr. Duckham will be a guide.

"Pure oleic acid is best prepared from tallow (which practically does not contain any less saturated acids than oleic) by saponifying with caustic potash, precipitating the soap solution with lead acetate, and extracting the dried lead salt with ether. The dissolved lead salt is decomposed with hydrochloric acid under ether, the liberated acid dissolved in ammonia, and the solution precipitated with barium chloride. Next the barium salt is dried, boiled out with hot alcohol, and the hot solution allowed to crystallize. The crystallized salt is decomposed either with strong mineral acid, or by tartaric acid. This is as far as the matter is ever carried to prepare pure oleic acid for the manufacture of ' pure ' sodium oleate, but for the most accurate chemical work it may be necessary to remove the remaining traces of solid acids, and this is done by the conversion into the chloro-iodide product, and crystallization from an organic solvent. The solid acids remain dissolved, and oleic acid can be obtained from the crystallized chloro-iodo product by heating with aniline."

I would only add to this a caution to the inexperienced that the lead salt referred to is oleate of lead and the solution of this in ether if spilt on the hand and not washed off at once will send the experimenter to bed with severe lead poisoning.

Rings for Bubbles

These may be made of any kind of wire. I have used tinned iron about one-twentieth of an inch in diameter. The joint should be smoothly soldered without lumps. If soldering is a difficulty, then use the thinnest wire that is stiff enough to support the bubbles steadily, and make the joint by twisting the end of the wire round two or three times. Rings two to three inches in diameter are convenient. I have seen that dipping the rings in melted paraffin is recommended, but I have not found any advantage from this. The nicest material for the light rings is thin aluminium wire, about as thick as a fine pin (No. 26 to 30 B.W.G.), and as this cannot be readily soldered, the ends must be twisted. If this is not to be had, very fine wire, nearly as fine as a hair (No. 36 B.W.G.), of copper or of any other metal, will answer. The rings should be wetted with the soap mixture before a bubble is placed upon them, and must always be well washed and dried when done with.

Threads in Ring

There is no difficulty in showing these experiments. The ring with the thread may be dipped in the soap solution, or stroked across with the edge of a piece of paper, celluloid or india-rubber sheet that has been dipped in the liquid, so as to form a film on both sides of the thread. A needle that has also been wetted with the soap may be used to show that the threads are loose. The same needle held for a moment in a candle-flame or a point of blotting paper supplies a convenient means of breaking the film.

Blow out Candle with Soap-bubble

For this, the bubble should be blown on the ends of a short wide pipe spread out at one end to give a better hold for the bubble. The tin funnel supplied with an ordinary gazogene answers perfectly. This should be washed before it is used again for filling the gazogene.

Bubbles Balanced against One Another

These experiments are most conveniently made on a small scale. Pieces of thin brass tube, three-eighths or half an inch in diameter, are suitable. It is best to have pieces of apparatus, specially prepared with taps, for easily and quickly stopping the air from leaving either bubble, and for putting the two bubbles into communication when required. It should not be difficult to contrive to perform the experiments, using india-rubber connecting tubes, pinched with spring clips to take the place of taps. There is one little detail which just makes the difference between success and failure. This is to supply a mouth piece for blowing the bubble, made of glass tube, which has been drawn out so fine that these little bubbles cannot be blown out suddenly by accident. It is very difficult, otherwise, to adjust the quantity of air in such small bubbles with any accuracy. In balancing a spherical against a cylindrical bubble, the short piece of tube, into which the air is supplied, must be made so that it can be easily moved to or from a fixed piece of the same size closed at the other end. Then the two ends of the short tube must have a film spread over them with a piece of paper, or india-rubber, but there must be *no* film stretched across the end of the fixed tube. The two tubes must at first be near together, until the spherical bubble has been formed. They may then be separated gradually more and more, and air blown in so as to keep the sides of the cylinder straight, until the cylinder is sufficiently long to be nearly unstable. It will then far more evidently show, by its change of form, than it would if it were short, when the pressure due to the spherical bubble exactly balances that due to a cylindrical one. If the shadow of the bubbles, or an image formed by a lens on a screen, is then measured, it will be found that the sphere has a diameter which is very accurately double that of the cylinder.

Water-drops in Paraffin and Bisulphide of Carbon

All that was said in describing the Plateau experiment applies here. Perfectly spherical and large drops of water can be formed in a mixture so made that the lower parts are very little heavier, and the upper parts very little lighter, than water. The addition of bisulphide of carbon makes the mixture heavier. This liquid—bisulphide of carbon—is very dangerous, and has a most dreadful smell, so that it had better not be brought into the house. The form of a hanging drop, and the way in which it breaks off, can be seen if water is used in paraffin alone, but it is much more evident if a little bisulphide of carbon is mixed with the paraffin, so that water will sink slowly in the mixture. Pieces of glass tube, open at both ends from half an inch to one inch in diameter, show the action best. Having poured some water coloured blue into a glass vessel, and covered it to a depth of several inches with paraffin, or the paraffin mixture, dip the pipe down into the water, having first closed the upper end with the thumb or the palm of the hand. On then removing the hand, the water will rush up inside the tube. Again close the upper end as before, and raise the tube until the lower end is well above the water, though still immersed in the paraffin. Then allow air to enter the pipe very slowly by just rolling the thumb the least bit to one side. The water will escape slowly and form a large growing drop, the size of which before it breaks away will depend on the density of the mixture and the size of the tube.

To form a water cylinder in the paraffin the tube must be filled with water as before, but the upper end must now be left open. Then when all is quiet the tube is to be rather rapidly withdrawn in the direction of its own length, when the water which was within it will be left behind in form of a cylinder, surrounded by the paraffin. It will then break up into spheres so slowly, in the case of a large tube, that the operation can be watched. The depth of paraffin should be quite ten times the diameter of the tube.

To make bubbles of water in the paraffin, the tube must be dipped down into the water with the upper end open all the time, so that the tube is mostly filled with paraffin. It must then be closed for a moment above and raised till the end is completely out of the water. Then if air is allowed to enter slowly, and the tube is gently raised, bubbles of water filled with paraffin will be formed which can be made to separate from the pipe, like soap-bubbles from a " churchwarden," by a suitable sudden movement. If a number of water-drops are floating in the paraffin in the pipe, and this can be easily arranged, then the bubbles made will contain possibly a number of other drops or even other bubbles. A very little bisulphide of carbon poured carefully down a pipe will form a heavy layer above the water, on which these compound bubbles will remain floating.

Cylindrical bubbles of water in paraffin may be made by dipping the pipe down into the water and withdrawing it quickly without ever closing the top at all. These break up into spherical bubbles in the same way that the cylinder of liquid broke up into spheres of liquid.

As before, orthotoluidine and water or salt water are more convenient than the bisulphide of carbon mixture and water.

Beaded Spider-webs

These are found in the spiral part of the webs of all the geometrical spiders. The beautiful geometrical webs may be found out of doors in abundance in the autumn, or in green-houses at almost any time of the year. To mount these webs so that the beads may be seen, take a small flat ring of any material, or a piece of cardboard with a hole cut out with a gun-wad cutter, or otherwise. Smear the face of the ring, or the card, with a very little strong gum. Choose a freshly-made web, and then pass the ring, or the card, across the web so that some of the spiral web (not the central part of the web) remains stretched across the hole. This must be done without touching or damaging the pieces that are stretched across, except at their ends. The beads are too small to be seen with the naked eye. A strong magnifying-glass, or a low-power microscope, will show the beads and their marvellous regularity. The beads on the webs of very young spiders are not so regular as those on spiders that are fully grown. Those beautiful beads, easily visible to the naked eye, on spider lines in the early morning of an autumn day, are not made by the spider, but are simply dew. They very perfectly show the spherical form of small water-drops.

Photographs of Water-jets

These are easily taken by the method described by Mr. Chichester Bell. The flash of light is produced by a short spark from a few Leyden-jars. The fountain, or jet, should be five or six feet away from the spark, and the photographic plate should be held as close to the stream of water as is possible without touching. The shadow is then so definite that the photograph, when taken, may be examined with a powerful lens, and will still appear sharp. Any rapid dry plate will do. The room, of course, must be quite dark when the plate is placed in position, and the spark then made. The regular breaking up of the jet may be effected by sound produced in almost any way. The straight jet, of which Fig. 41 is a representation, magnified about three and a quarter times, was regularly broken up by simply whistling to it with a key. The fountains were broken up regularly by fastening the nozzle to one end of a long piece of wood clamped at the end to the stand of a tuning-fork, which was kept sounding by electrical means. An ordinary tuning-fork, made to rest when sounding against the wooden support of the nozzle, will answer quite as well, but it is not quite so convenient. The jet will break up best to certain notes, but it may be tuned to a great extent by altering the size of the orifice or the pressure of the water, or both.

Fountain and Sealing-wax

It is almost impossible to fail over this very striking yet simple experiment. A fountain of almost any size, at any rate between one-fiftieth and a quarter of an inch in the smooth part, and up to eight feet high, will cease to scatter when the sealing-wax is rubbed with flannel and held a few feet away. A suitable size of fountain is one about four feet high, coming from an orifice anywhere near one-sixteenth of an inch in diameter. The nozzle should be inclined so that the water falls slightly on one side. The sealing-wax may be electrified by being rubbed on the coat-sleeve, or on a piece of fur or flannel which is *dry*. It will then make little pieces of paper or cork dance, but it will still act on the fountain when it has ceased to produce any visible effect on pieces of paper, or even on a delicate gold-leaf electroscope.

Bouncing Water-jets

This beautiful experiment of Lord Rayleigh's requires a little management to make it work in a satisfactory manner. Take a piece of quill-glass tube and draw it out to a very slight extent (see a former note), so as to make a neck about one-eighth of an inch in diameter at the narrowest part. Break the tube just at this place, after first nicking it there with a file. Connect each of these tubes by means of an india-rubber pipe, or otherwise, with a supply of water in a bottle, and pinch the tubes with a screw-clip until two equal jets of water are formed. So hold the nozzles that these meet in their smooth portions at a very small angle. They will then for a short time bounce away from one another without mixing. If the air is very dusty, if the water is not clean, or if air-bubbles are carried along in the pipes, the two jets will at once join together. In the arrangement that I used in the lantern, the two nozzles were nearly horizontal, one was about half an inch above the other, and they were very slightly converging. They were fastened in their position by melting upon them a little sealing-wax. India-rubber pipes connected them with two bottles about six inches above them, and screw-clips were used to regulate the supply. One of the bottles was made to stand on three pieces of sealing-wax to insulate it electrically, and the corresponding nozzle was only held by its sealing-wax fastening. The water in the bottles had been filtered, and one was coloured blue. If these precautions are taken, the jets will remain distinct quite long enough, but are instantly caused to recombine by a piece of electrified sealing-wax six or eight feet away. They may be separated again by touching the water issuing near one nozzle with the finger, which deflects it ; on quietly removing the finger the jet takes up its old position and bounces off the other as before. They can thus be separated and made to combine ten or a dozen times in a minute.

Fountain and Intermittent Light

This can be successfully shown to a large number of people at once only by using an electric arc, but there is no occasion to produce this light if not more than one person at a time wishes to see the evolution of the drops. It is then merely necessary to make the fountain play in front of a bright background such as the sky, to break it up with a tuning-fork or other musical sound as described, and then to look at it through a card disc equally divided near the edge into spaces about two or three inches wide, with a hole about one-eighth of an inch in diameter between each pair of spaces. A disc of card five inches in diameter, with six equidistant holes half an inch from the edge, answers well. The disc must be made to spin by any means very regularly at such a speed that the tuning-fork, or stretched string if this be used, when looked at through the holes, appears quiet, or nearly quiet, when made to vibrate. The separate drops will then be seen, and everything described in the preceding pages, and a great deal more, will be evident. This is one of the most fascinating experiments, and it is well worth while to make an effort to succeed. The card is most conveniently carried by a small electric motor driven by a few secondary cells and regulated by a resistance to run just too fast, then the speed can be perfectly regulated by a very light pressure of the finger on the end of the axle.

Mr. Chichester Bell's Singing Water-jet

For these experiments a very fine hole about one seventy-fifth of an inch in diameter is most suitable. To obtain this, Mr. Bell holds the end of a quill-glass tube in a blow-pipe flame, and constantly turns it round and round until the end is almost entirely closed up. He then suddenly and forcibly blows into the pipe. Out of several nozzles made in this way, some are sure to do well. Lord Rayleigh makes nozzles generally by cementing to the end of a glass (or metal) pipe a piece of thin sheet metal in which a hole of the required size has been made, e. g. by a punch made from a broken needle on a block of lead. The water pressure should be produced by a head of about fifteen feet. The water must be quite free from dust and from air-bubbles. This may be effected by making it pass through a piece of tube stuffed full of flannel, or cotton-wool, or something of the kind to act as a filter. There should be a yard or so of good black india-rubber tube, about one-eighth of an inch in diameter inside between the filter and the nozzle. It is best not to take the water direct from the water-main, but from a cistern about fifteen feet above the nozzle. If no cistern is available, a pail of water taken up-stairs, with a pipe coming down, is an excellent substitute, and this has the further advantage that the head of water can be easily changed so as to arrive at the best result.

The rest of the apparatus is very simple. It is merely necessary to stretch and tie over the end of a tube about half an inch in diameter a piece of thin india-rubber sheet, cut from an air-ball that has not been blown out. The tube, which may be of metal or of glass, may either be fastened to a heavy foot, in which case a side tube must be joined to it, as in Fig. 47, or it may be open at both ends and be held in a clamp. It is well to put a cone of cardboard on the open end (Fig. 48), if the sound is to be heard by many at a time. If the experimenter alone wishes to hear as well as possible when faint sounds are produced, he should carry a piece of smooth india-rubber tube about half an inch in diameter from the open end to his ear. This, however, would nearly deafen him with such loud noises as the tick of a watch.

Bubbles and Ether

Experiments with ether must be performed with great care, because, like the bisulphide of carbon, it is dangerously inflammable. The bottle of ether must never be brought near a light. If a large quantity is spilled, the heavy vapour is apt to run along the floor and ignite at a fire, even on the other side of a room. Any vessel may be filled with the vapour of ether by merely pouring the liquid upon a piece of blotting-paper reaching up to the level of the edge. Very little is required, say half a wine-glassful, for a basin that would hold a gallon or more. In a draughty place the vapour will be lost in a short time. Bubbles can be set to float upon the vapour without any difficulty. They may be removed in five or ten seconds by means of one of the small light rings with a handle, provided that the ring is wetted with the soap solution and has *no* film stretched across it. If taken to a light at a safe distance the bubble will immediately burst into a blaze. If a neighbouring light is not close down to the table, but well up above the jar on a stand, it may be near with but little risk. To show the burning vapour, the same wide tube

that was used to blow out the candle will answer well.
The pear shape of the bubble, owing to its increased
weight after being held in the vapour for ten or fifteen
seconds, is evident enough on its removal, but the
falling stream of heavy vapour, which comes out again
afterwards, can only be shown if its shadow is cast upon
a screen by means of a bright light.

Experiment with Internal Bubbles

For these experiments, next to a good solution, the
pipe is of the greatest importance. A "churchwarden"
is no use. A glass pipe $\frac{5}{16}$ inch in diameter at the
mouth is best. If this is merely a tube bent near the
end through a right angle, moisture condensed in the
tube will in time run down and destroy the bubble
occasionally, which is very annoying in a difficult
experiment. I have made for myself the pipe of which
Fig. 81 is a full-size representation, and I do not think
that it is possible to improve upon this. Those who are
not glass-blowers will be able, with the help of cork, to
make a pipe with a trap as shown in Fig. 82, which is as
good, except in appearance and handiness.

In knocking bubbles together to show that they do
not touch, care must be taken to avoid letting either
bubble meet any projection in the other, such as the
wire ring, or a heavy drop of liquid. Either will instantly
destroy the two bubbles. There is also a limit to the
violence which may be used, which experience will soon
indicate.

In pushing a bubble through a ring smaller than itself,
by means of a flat film on another ring, it is important
that the bubble should not be too large; but a larger
bubble can be pushed through than would be expected.
It is not so easy to push it up as down because of the
heavy drop of liquid, which it is difficult to drain away
completely.

To blow one bubble inside another, the first, as large as an average orange, should be blown on the lower side

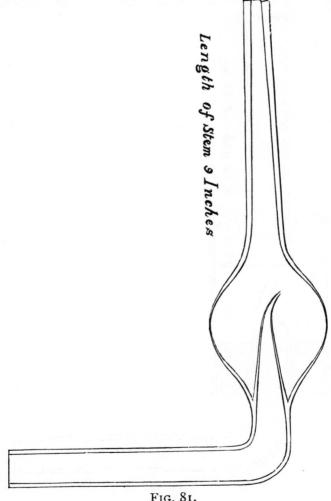

Length of Stem 9 Inches

FIG. 81.

of a horizontal ring. A light wire ring should then be hung on to this bubble to pull it slightly out of shape. For this purpose thin aluminium rings are hardly heavy enough, and so either a heavier metal should be used, or a small weight should be fastened to the handle of the ring. The ring should be so heavy that the sides of the

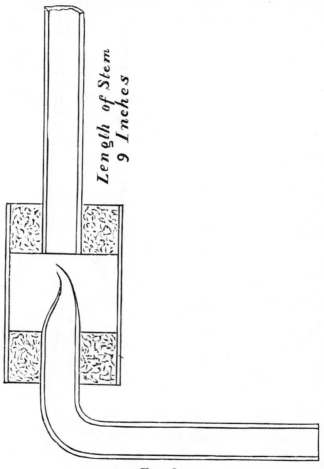

Length of Stem 9 Inches

FIG. 82.

bubble make an angle of thirty or forty degrees with the vertical, where they meet the ring as indicated in Fig. 57. The wetted end of the pipe is now to be inserted through the top of the bubble, until it has penetrated a clear half inch or so. A new bubble can now be blown any size almost that may be desired. To remove the pipe a slow motion will be fatal, because it will raise the inner bubble until it and the outer one both meet the pipe at the same place. This will bring them into true contact. On the other hand, a violent jerk will almost certainly produce too great a disturbance. A rather rapid motion, or a slight jerk, is all that is required. It is advisable before passing the pipe up through the lower ring, so as to touch the inner bubble, and so drain away the heavy drop, to steady the ring with the other hand. The superfluous liquid can then be drained from both bubbles simultaneously. Care must be taken after this that the inner bubble is not allowed to come against either wire ring, nor must the pipe be passed through the side where the two bubbles are very close together. To peel off the lower ring it should be pulled down a very little way and then inclined to one side. The peeling will then start more readily, but as soon as it has begun the ring should be raised so as not to make the peeling too rapid, otherwise the final jerk, when it leaves the lower ring, will be too much for the bubbles to withstand.

Bubbles coloured with fluorescine, or uranine, do not show their brilliant fluorescence unless sunlight or electric light is concentrated upon them with a lens or mirror. The quantity of dye required is so small that it may be difficult to take little enough. As much as can be picked up on the last eighth of an inch of a pointed pen-knife will be, roughly speaking, enough for a wine-glassful of the soap solution. If the quantity is increased beyond something like the proportion stated, the fluorescence becomes less and very soon disappears. The best quantity can be found in a few minutes by trial.

To blow bubbles containing either coal-gas or air, or a mixture of the two, the most convenient plan is to have a small T-shaped glass tube which can be joined by one arm of the T to the blow-pipe by means of a short piece of india-rubber tube, and be connected by its vertical limb with a sufficient length of india-rubber pipe, one eighth of an inch in diameter inside, to reach to the floor, after which it may be connected by any kind of pipe with the gas supply. The gas can be stopped either by pinching the india-rubber tube with the left hand, if that is at liberty, or by treading on it if both hands are occupied. Meanwhile air can be blown in by the other arm of the T, and the end closed by the tongue when gas alone is required. This end of the tube should be slightly spread out when hot by rapidly pushing into it the *cold* tang of a file, and twisting it at the same time, so that it may be lightly held by the teeth without fear of slipping.

If a light T-piece or so great a length of small india-rubber tube cannot be obtained, then the mouth must be removed from the pipe and the india-rubber tube slipped in when air is to be changed for gas. This makes the manipulation more difficult, but all the experiments, except the one with three bubbles can be so carried out.

The pipe must in every case be made to enter the highest point of a bubble in order to start an internal one. If it is pushed horizontally through the side, the inner bubble is sure to break. If the inner bubble is being blown with gas, it will soon tend to rise. The pipe must then be turned over in such a manner that the inner bubble does not creep along it, and so meet the outer one where penetrated by the pipe. A few trials will show what is meant. The inner bubble may then be allowed to rest against the top of the outer one while being enlarged. When it is desired after withdrawing the pipe to blow more air or gas into either the inner or the outer bubble, it is not safe after inserting the pipe again to begin to blow at once ; the film which is now stretched across the mouth of the pipe will probably become a third bubble, and this, in the circumstances, is almost certain to cause a failure. An instantaneous withdrawal of the air destroys this film by drawing it into the pipe. Air or gas may then be blown without danger.

If the same experiment is performed upon a light ring with cotton and paper attached, the left hand will be occupied in holding this ring, and then the gas must be controlled by the foot, or by a friend. The light ring should be quite two inches in diameter. If, when the inner bubble has begun to carry away the ring, &c., the paper is caught hold of, it is possible, by a judicious pull, to cause the two bubbles to leave the ring and so escape into the air one inside the other. For this purpose the smallest ring that will carry the paper should be used. With larger rings the same effect may be produced by inclining the ring, and so allowing the outer bubble to peel off, or by placing the mouth of the pipe against the ring and blowing a third bubble in real contact with the ring and the outer bubble. This will assist the peeling process.

To blow three bubbles, one inside the other two, is more difficult. The following plan I have found to be fairly certain. First blow above the ring a bubble the size of a large orange. Then take a small ring about an inch in diameter, with a straight wire coming down from one side to act as a handle, and after wetting it with the solution, pass it carefully up through the fixed ring so that the small ring is held well inside the bubble. Now pass the pipe, freshly dipped in the solution, into the outer or No. 1 bubble until it is quite close to the small ring, and begin to blow the No. 2 bubble. This must be started with the pipe almost in contact with the inner ring, as the film on this ring would destroy a bubble that had attained any size. Withdraw the pipe, dip it into the liquid, and insert it into the inner bubble, taking care to keep these two bubbles from meeting anywhere. Now blow a large gas-bubble, which may rest against the top of No. 2 while it is growing. No. 2 may now rest against

the top of No. 1 without danger. Remove the pipe from
No. 3 by gently lowering it, but without any gas passing
at the time, and then let some gas into No. 2 to make
it lighter, and at the same time diminish the pressure
between Nos. 2 and 3. Presently the small ring can be
peeled off No. 2 and removed altogether. But if there is
a difficulty in accomplishing this, withdraw the pipe from
No. 2 and blow air into No. 1 to enlarge it, which will
make the process easier. Then remove the pipe from
No. 1. The three bubbles are now resting one inside the
other. By blowing a fourth bubble, as described above,
against the fixed ring, No. 1 bubble will peel off, and the
three will float away. No. 1 can, while peeling, be
transferred to a light wire ring from which paper, &c., are
suspended. This description sounds complicated, but
after a little practice the process can be carried out
almost with certainty in far less time than it takes to
describe it ; in fact, so quickly can it be done, and so
simple does it appear, that no one would suppose that
so many details had to be attended to.

Bubbles and Electricity

These experiments are on the whole the most difficult
to perform successfully. The following details should be
sufficient to prevent failure. Two rings are formed at the
end of a pair of wires about six inches long in the straight
part. About one inch at the opposite end from the ring
is turned down at a right angle. These turned-down
ends rest in two holes drilled vertically in a non-conductor
such as ebonite, about two or three inches apart. Then
if all is right the two rings are horizontal and at the same
level, and they may be moved towards or away from one
another. Separate them a few inches, and blow a bubble
above or below each, making them nearly the same size.
Then bring the two rings nearer together until the
bubbles just, and only just, rest against one another.

Though they may be hammered together without joining, they will not remain long resting in this position, as the convex surfaces can readily squeeze out the air. The ebonite should not be perfectly warm and dry, for it is then sure to be electrified, and this will give trouble. It must not be wet, because then it will conduct, and the sealing-wax will produce no result. If it has been used as the support for the rings for some of the previous experiments, it will have been sufficiently splashed by the bursting of bubbles to be in the best condition. It must, however, be well wiped occasionally.

A stick of sealing-wax should be held in readiness under the arm, in a fold or two of *dry* flannel or fur. If the wax is very strongly electrified, it is apt to be far too powerful, and to cause the bubbles, when it is presented to them, to destroy each other. A feeble electrification is sufficient; then the instant it is exposed the bubbles coalesce. The wax may be brought so near one bubble in which another one is resting, that it pulls them to one side, but the inner one is screened from electrical action by the outer one. It is important not to bring the wax very near, as in that case the bubble will be pulled so far as to touch it, and so be broken. The wetting of the wax will make further electrification very uncertain. In showing the difference between an inner and an outer bubble, the same remarks with regard to undue pressure, electrification or loss of time apply. I have generally found that it is advisable in this experiment not to drain the drops from both the bubbles, as their weight seems to steady them; the external bubble may be drained, and if it is not too large, the process of electrically joining the outer bubbles, without injury to the inner one, may be repeated many times. I once caused eight or nine single bubbles to unite with the outer one of a pair in succession before it became too unwieldy for more accessions to be possible.

It would be going outside my subject to say anything about the management of lanterns. I may, however, state that while the experiments with the small bubbles are best projected with a lens upon the screen, the larger bubbles described in the last lecture can only be projected by their shadows. For this purpose the condensing lens is removed, and the bare light alone made use of. An electric arc is far preferable to a lime-light, both because the shadows are sharper, and because the colours are so much more brilliant. No oil lamp would answer, even if the light were sufficient in quantity, because the flame would be far too large to cast a sharp shadow. A Nernst lamp answers fairly well, but limelight is better.

In these hints, which have in themselves required a rather formidable chapter, I have given all the details, so far as I am able, which a considerable experience has shown to be necessary for the successful performance of the experiments in public. The hints will I hope materially assist those who are not in the habit of carrying out experiments, but who may wish to perform them for their own satisfaction. Though people who are not experimentalists may consider that the hints are overburdened with detail, it is probable that in repeating the experiments they will find here and there, in spite of all my care to provide against unforeseen difficulties, that more detail would have been desirable.

Thaumatrope for showing the Formation and Oscillations of Drops.

The experiment showing the formation of water-drops can be very perfectly imitated, and the movements actually made visible, without any necessity for using liquids at all, by simply converting Fig. 83 (at end of book) into the old-fashioned instrument called a thaumatrope. What will then be seen is a true representation, because the forms in the figure are copies of a series of photographs taken from the moving drops at the rate of forty-three photographs in two seconds.[1]

Obtain a piece of good cardboard as large as the figure, and having brushed it all over on one side with thin paste, lay the figure upon it, and press it down evenly. Place it upon a table, and cover it with a few thicknesses of blotting-paper, and lay over all a flat piece of board large enough to cover it. Weights sufficient to keep it all flat may be added. This must be left all night at least, until the card is quite dry, or else it will curl up and be useless. Now with a sharp chisel or knife, but a chisel if possible, cut out the forty-three slits near the edge, accurately following the outline indicated in black and white, and keeping the slits as narrow as possible. Then cut a hole in the middle, so as to fit the projecting part of a sewing-machine cotton-reel, and fasten the cotton-reel on the side away from the figure with glue or small nails. It must be fixed exactly in the middle. The edge should of course be cut down to the outside of the black rim.

Now having found a pencil or other rod on which the cotton-reel will freely turn, use this as an axle, and holding the disc up in front of a looking-glass, and in a good light, slowly and steadily make it turn round. The image of the disc seen through the slit in the looking-glass will then perfectly represent every feature of the growing and falling drop. As the drop grows it will gradually become too heavy to be supported, a waist will then begin to form which will rapidly get narrower, until the drop at last breaks away. It will be seen to continue its fall until it has disappeared in the liquid below, but it has not mixed with this, and so it will presently appear again, having bounced out of the liquid. As it falls it will be seen to vibrate

[1] For particulars see *Philosophical Magazine*, September 1890.

as the result of the sudden release from the one-sided pull. The neck which was drawn out will meanwhile have gathered itself in the form of a little drop, which will then be violently hit by the oscillations of the remaining pendant drop above, and driven down. The pendant drop will be seen to vibrate and grow at the same time, until it again breaks away as before, and so the phenomena are repeated.

In order to perfectly reproduce the experiment, the axle should be firmly held upon a stand, and the speed should not exceed one turn in two seconds.

The effect is still more real if a screen is placed between the disc and the mirror, which will only allow one of the drops to be seen.

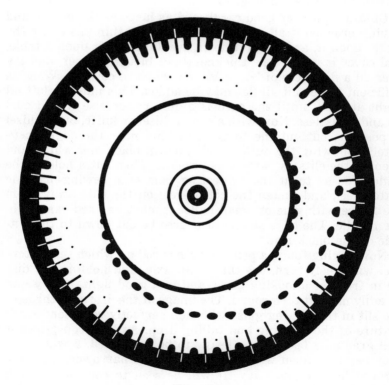

Fig. 83

THAUMATROPE for showing

the formation and oscillation of drops.

A CATALOG OF SELECTED DOVER BOOKS IN ALL FIELDS OF INTEREST

CONCERNING THE SPIRITUAL IN ART, Wassily Kandinsky. Pioneering work by father of abstract art. Thoughts on color theory, nature of art. Analysis of earlier masters. 12 illustrations. 80pp. of text. 5⅜ × 8½. 23411-8 Pa. $2.95

LEONARDO ON THE HUMAN BODY, Leonardo da Vinci. More than 1200 of Leonardo's anatomical drawings on 215 plates. Leonardo's text, which accompanies the drawings, has been translated into English. 506pp. 8⅜ × 11¾. 24483-0 Pa. $11.95

GOBLIN MARKET, Christina Rossetti. Best-known work by poet comparable to Emily Dickinson, Alfred Tennyson. With 46 delightfully grotesque illustrations by Laurence Housman. 64pp. 4 × 6¾. 24516-0 Pa. $2.50

THE HEART OF THOREAU'S JOURNALS, edited by Odell Shepard. Selections from *Journal*, ranging over full gamut of interests. 228pp. 5⅜ × 8½. 20741-2 Pa. $4.50

MR. LINCOLN'S CAMERA MAN: MATHEW B. BRADY, Roy Meredith. Over 300 Brady photos reproduced directly from original negatives, photos. Lively commentary. 368pp. 8⅜ × 11¼. 23021-X Pa. $14.95

PHOTOGRAPHIC VIEWS OF SHERMAN'S CAMPAIGN, George N. Barnard. Reprint of landmark 1866 volume with 61 plates: battlefield of New Hope Church, the Etawah Bridge, the capture of Atlanta, etc. 80pp. 9 × 12. 23445-2 Pa. $6.00

A SHORT HISTORY OF ANATOMY AND PHYSIOLOGY FROM THE GREEKS TO HARVEY, Dr. Charles Singer. Thoroughly engrossing non-technical survey. 270 illustrations. 211pp. 5⅜ × 8½. 20389-1 Pa. $4.95

REDOUTE ROSES IRON-ON TRANSFER PATTERNS, Barbara Christopher. Redouté was botanical painter to the Empress Josephine; transfer his famous roses onto fabric with these 24 transfer patterns. 80pp. 8¼ × 10⅞. 24292-7 Pa. $3.50

THE FIVE BOOKS OF ARCHITECTURE, Sebastiano Serlio. Architectural milestone, first (1611) English translation of Renaissance classic. Unabridged reproduction of original edition includes over 300 woodcut illustrations. 416pp. 9⅜ × 12¼. 24349-4 Pa. $14.95

CARLSON'S GUIDE TO LANDSCAPE PAINTING, John F. Carlson. Authoritative, comprehensive guide covers, every aspect of landscape painting. 34 reproductions of paintings by author; 58 explanatory diagrams. 144pp. 8⅜ × 11. 22927-0 Pa. $5.95

101 PUZZLES IN THOUGHT AND LOGIC, C.R. Wylie, Jr. Solve murders, robberies, see which fishermen are liars—purely by reasoning! 107pp. 5⅜ × 8½. 20367-0 Pa. $2.00

TEST YOUR LOGIC, George J. Summers. 50 more truly new puzzles with new turns of thought, new subtleties of inference. 100pp. 5⅜ × 8½. 22877-0 Pa. $2.50

THE MURDER BOOK OF J.G. REEDER, Edgar Wallace. Eight suspenseful stories by bestselling mystery writer of 20s and 30s. Features the donnish Mr. J.G. Reeder of Public Prosecutor's Office. 128pp. 5⅜ × 8½.

24374-5 Pa. $3.95

ANNE ORR'S CHARTED DESIGNS, Anne Orr. Best designs by premier needlework designer, all on charts: flowers, borders, birds, children, alphabets, etc. Over 100 charts, 10 in color. Total of 40pp. 8¼ × 11. 23704-4 Pa. $2.50

BASIC CONSTRUCTION TECHNIQUES FOR HOUSES AND SMALL BUILDINGS SIMPLY EXPLAINED, U.S. Bureau of Naval Personnel. Grading, masonry, woodworking, floor and wall framing, roof framing, plastering, tile setting, much more. Over 675 illustrations. 568pp. 6½ × 9¼. 20242-9 Pa. $9.95

MATISSE LINE DRAWINGS AND PRINTS, Henri Matisse. Representative collection of female nudes, faces, still lifes, experimental works, etc., from 1898 to 1948. 50 illustrations. 48pp. 8⅜ × 11¼. 23877-6 Pa. $3.50

HOW TO PLAY THE CHESS OPENINGS, Eugene Znosko-Borovsky. Clear, profound examinations of just what each opening is intended to do and how opponent can counter. Many sample games. 147pp. 5⅜ × 8½. 22795-2 Pa. $3.50

DUPLICATE BRIDGE, Alfred Sheinwold. Clear, thorough, easily followed account: rules, etiquette, scoring, strategy, bidding; Goren's point-count system, Blackwood and Gerber conventions, etc. 158pp. 5⅜ × 8½. 22741-3 Pa. $3.50

SARGENT PORTRAIT DRAWINGS, J.S. Sargent. Collection of 42 portraits reveals technical skill and intuitive eye of noted American portrait painter, John Singer Sargent. 48pp. 8¼ × 11⅛. 24524-1 Pa. $3.50

ENTERTAINING SCIENCE EXPERIMENTS WITH EVERYDAY OBJECTS, Martin Gardner. Over 100 experiments for youngsters. Will amuse, astonish, teach, and entertain. Over 100 illustrations. 127pp. 5⅜ × 8½. 24201-3 Pa. $2.50

TEDDY BEAR PAPER DOLLS IN FULL COLOR: A Family of Four Bears and Their Costumes, Crystal Collins. A family of four Teddy Bear paper dolls and nearly 60 cut-out costumes. Full color, printed one side only. 32pp. 9¼ × 12¼.

24550-0 Pa. $3.50

NEW CALLIGRAPHIC ORNAMENTS AND FLOURISHES, Arthur Baker. Unusual, multi-useable material: arrows, pointing hands, brackets and frames, ovals, swirls, birds, etc. Nearly 700 illustrations. 80pp. 8⅜ × 11¼.

24095-9 Pa. $3.75

DINOSAUR DIORAMAS TO CUT & ASSEMBLE, M. Kalmenoff. Two complete three-dimensional scenes in full color, with 31 cut-out animals and plants. Excellent educational toy for youngsters. Instructions; 2 assembly diagrams. 32pp. 9¼ × 12¼. 24541-1 Pa. $4.50

SILHOUETTES: A PICTORIAL ARCHIVE OF VARIED ILLUSTRATIONS, edited by Carol Belanger Grafton. Over 600 silhouettes from the 18th to 20th centuries. Profiles and full figures of men, women, children, birds, animals, groups and scenes, nature, ships, an alphabet. 144pp. 8⅜ × 11¼. 23781-8 Pa. $5.95

CATALOG OF DOVER BOOKS

25 KITES THAT FLY, Leslie Hunt. Full, easy-to-follow instructions for kites made from inexpensive materials. Many novelties. 70 illustrations. 110pp. 5⅜ × 8½.
22550-X Pa. $2.50

PIANO TUNING, J. Cree Fischer. Clearest, best book for beginner, amateur. Simple repairs, raising dropped notes, tuning by easy method of flattened fifths. No previous skills needed. 4 illustrations. 201pp. 5⅜ × 8½.   23267-0 Pa. $3.50

EARLY AMERICAN IRON-ON TRANSFER PATTERNS, edited by Rita Weiss. 75 designs, borders, alphabets, from traditional American sources. 48pp. 8¼ × 11.
23162-3 Pa. $1.95

CROCHETING EDGINGS, edited by Rita Weiss. Over 100 of the best designs for these lovely trims for a host of household items. Complete instructions, illustrations. 48pp. 8¼ × 11.   24031-2 Pa. $2.95

FINGER PLAYS FOR NURSERY AND KINDERGARTEN, Emilie Poulsson. 18 finger plays with music (voice and piano); entertaining, instructive. Counting, nature lore, etc. Victorian classic. 53 illustrations. 80pp. 6½ × 9¼. 22588-7 Pa. $2.25

BOSTON THEN AND NOW, Peter Vanderwarker. Here in 59 side-by-side views are photographic documentations of the city's past and present. 119 photographs. Full captions. 122pp. 8¼ × 11.   24312-5 Pa. $7.95

CROCHETING BEDSPREADS, edited by Rita Weiss. 22 patterns, originally published in three instruction books 1939-41. 39 photos, 8 charts. Instructions. 48pp. 8¼ × 11.   23610-2 Pa. $2.00

HAWTHORNE ON PAINTING, Charles W. Hawthorne. Collected from notes taken by students at famous Cape Cod School; hundreds of direct, personal *apercus*, ideas, suggestions. 91pp. 5⅜ × 8½.   20653-X Pa. $2.95

THERMODYNAMICS, Enrico Fermi. A classic of modern science. Clear, organized treatment of systems, first and second laws, entropy, thermodynamic potentials, etc. Calculus required. 160pp. 5⅜ × 8½.   60361-X Pa. $4.50

TEN BOOKS ON ARCHITECTURE, Vitruvius. The most important book ever written on architecture. Early Roman aesthetics, technology, classical orders, site selection, all other aspects. Morgan translation. 331pp. 5⅜ × 8½. 20645-9 Pa. $6.95

THE CORNELL BREAD BOOK, Clive M. McCay and Jeanette B. McCay. Famed high-protein recipe incorporated into breads, rolls, buns, coffee cakes, pizza, pie crusts, more. Nearly 50 illustrations. 48pp. 8¼ × 11.   23995-0 Pa. $2.00

THE CRAFTSMAN'S HANDBOOK, Cennino Cennini. 15th-century handbook, school of Giotto, explains applying gold, silver leaf; gesso; fresco painting, grinding pigments, etc. 142pp. 6⅛ × 9¼.   20054-X Pa. $3.95

FRANK LLOYD WRIGHT'S FALLINGWATER, Donald Hoffmann. Full story of Wright's masterwork at Bear Run, Pa. 100 photographs of site, construction, and details of completed structure. 112pp. 9¼ × 10.   23671-4 Pa. $7.95

OVAL STAINED GLASS PATTERN BOOK, C. Eaton. 60 new designs framed in shape of an oval. Greater complexity, challenge with sinuous cats, birds, mandalas framed in antique shape. 64pp. 8¼ × 11.   24519-5 Pa. $3.95

THE BOOK OF WOOD CARVING, Charles Marshall Sayers. Still finest book for beginning student. Fundamentals, technique; gives 34 designs, over 34 projects for panels, bookends, mirrors, etc. 33 photos. 118pp. 7¾ × 10⅝.   23654-4 Pa. $3.95

CARVING COUNTRY CHARACTERS, Bill Higginbotham. Expert advice for beginning, advanced carvers on materials, techniques for creating 18 projects— mirthful panorama of American characters. 105 illustrations. 80pp. 8⅜ × 11.
24135-1 Pa. $2.95

300 ART NOUVEAU DESIGNS AND MOTIFS IN FULL COLOR, C.B. Grafton. 44 full-page plates display swirling lines and muted colors typical of Art Nouveau. Borders, frames, panels, cartouches, dingbats, etc. 48pp. 9⅜ × 12¼.
24354-0 Pa. $6.95

SELF-WORKING CARD TRICKS, Karl Fulves. Editor of *Pallbearer* offers 72 tricks that work automatically through nature of card deck. No sleight of hand needed. Often spectacular. 42 illustrations. 113pp. 5⅜ × 8½.   23334-0 Pa. $3.50

CUT AND ASSEMBLE A WESTERN FRONTIER TOWN, Edmund V. Gillon, Jr. Ten authentic full-color buildings on heavy cardboard stock in H-O scale. Sheriff's Office and Jail, Saloon, Wells Fargo, Opera House, others. 48pp. 9¼ × 12¼.
23736-2 Pa. $4.95

CUT AND ASSEMBLE AN EARLY NEW ENGLAND VILLAGE, Edmund V. Gillon, Jr. Printed in full color on heavy cardboard stock. 12 authentic buildings in H-O scale: Adams home in Quincy, Mass., Oliver Wight house in Sturbridge, smithy, store, church, others. 48pp. 9¼ × 12¼.   23536-X Pa. $4.95

THE TALE OF TWO BAD MICE, Beatrix Potter. Tom Thumb and Hunca Munca squeeze out of their hole and go exploring. 27 full-color Potter illustrations. 59pp. 4¼ × 5½. (Available in U.S. only)   23065-1 Pa. $1.75

CARVING FIGURE CARICATURES IN THE OZARK STYLE, Harold L. Enlow. Instructions and illustrations for ten delightful projects, plus general carving instructions. 22 drawings and 47 photographs altogether. 39pp. 8⅜ × 11.
23151-8 Pa. $2.95

A TREASURY OF FLOWER DESIGNS FOR ARTISTS, EMBROIDERERS AND CRAFTSMEN, Susan Gaber. 100 garden favorites lushly rendered by artist for artists, craftsmen, needleworkers. Many form frames, borders. 80pp. 8¼ × 11.
24096-7 Pa. $3.95

CUT & ASSEMBLE A TOY THEATER/THE NUTCRACKER BALLET, Tom Tierney. Model of a complete, full-color production of Tchaikovsky's classic. 6 backdrops, dozens of characters, familiar dance sequences. 32pp. 9⅜ × 12¼.
24194-7 Pa. $4.50

ANIMALS: 1,419 COPYRIGHT-FREE ILLUSTRATIONS OF MAMMALS, BIRDS, FISH, INSECTS, ETC., edited by Jim Harter. Clear wood engravings present, in extremely lifelike poses, over 1,000 species of animals. 284pp. 9 × 12.
23766-4 Pa. $9.95

MORE HAND SHADOWS, Henry Bursill. For those at their 'finger ends,'' 16 more effects—Shakespeare, a hare, a squirrel, Mr. Punch, and twelve more—each explained by a full-page illustration. Considerable period charm. 30pp. 6½ × 9¼.
21384-6 Pa. $1.95

SURREAL STICKERS AND UNREAL STAMPS, William Rowe. 224 haunting, hilarious stamps on gummed, perforated stock, with images of elephants, geisha girls, George Washington, etc. 16pp. one side. 8¼ × 11.   24371-0 Pa. $3.50

GOURMET KITCHEN LABELS, Ed Sibbett, Jr. 112 full-color labels (4 copies each of 28 designs). Fruit, bread, other culinary motifs. Gummed and perforated. 16pp. 8¼ × 11.   24087-8 Pa. $2.95

PATTERNS AND INSTRUCTIONS FOR CARVING AUTHENTIC BIRDS, H.D. Green. Detailed instructions, 27 diagrams, 85 photographs for carving 15 species of birds so life-like, they'll seem ready to fly! 8¼ × 11.   24222-6 Pa. $3.00

FLATLAND, E.A. Abbott. Science-fiction classic explores life of 2-D being in 3-D world. 16 illustrations. 103pp. 5⅜ × 8.   20001-9 Pa. $2.00

DRIED FLOWERS, Sarah Whitlock and Martha Rankin. Concise, clear, practical guide to dehydration, glycerinizing, pressing plant material, and more. Covers use of silica gel. 12 drawings. 32pp. 5⅜ × 8½.   21802-3 Pa. $1.00

EASY-TO-MAKE CANDLES, Gary V. Guy. Learn how easy it is to make all kinds of decorative candles. Step-by-step instructions. 82 illustrations. 48pp. 8¼ × 11.
23881-4 Pa. $2.95

SUPER STICKERS FOR KIDS, Carolyn Bracken. 128 gummed and perforated full-color stickers: GIRL WANTED, KEEP OUT, BORED OF EDUCATION, X-RATED, COMBAT ZONE, many others. 16pp. 8¼ × 11.   24092-4 Pa. $3.50

CUT AND COLOR PAPER MASKS, Michael Grater. Clowns, animals, funny faces...simply color them in, cut them out, and put them together, and you have 9 paper masks to play with and enjoy. 32pp. 8¼ × 11.   23171-2 Pa. $2.95

A CHRISTMAS CAROL: THE ORIGINAL MANUSCRIPT, Charles Dickens. Clear facsimile of Dickens manuscript, on facing pages with final printed text. 8 illustrations by John Leech, 4 in color on covers. 144pp. 8⅜ × 11¼.
20980-6 Pa. $5.95

CARVING SHOREBIRDS, Harry V. Shourds & Anthony Hillman. 16 full-size patterns (all double-page spreads) for 19 North American shorebirds with step-by-step instructions. 72pp. 9¼ × 12¼.   24287-0 Pa. $5.95

THE GENTLE ART OF MATHEMATICS, Dan Pedoe. Mathematical games, probability, the question of infinity, topology, how the laws of algebra work, problems of irrational numbers, and more. 42 figures. 143pp. 5⅜ × 8½.
22949-1 Pa. $3.50

READY-TO-USE DOLLHOUSE WALLPAPER, Katzenbach & Warren, Inc. Stripe, 2 floral stripes, 2 allover florals, polka dot; all in full color. 4 sheets (350 sq. in.) of each, enough for average room. 48pp. 8¼ × 11.   23495-9 Pa. $2.95

MINIATURE IRON-ON TRANSFER PATTERNS FOR DOLLHOUSES, DOLLS, AND SMALL PROJECTS, Rita Weiss and Frank Fontana. Over 100 miniature patterns: rugs, bedspreads, quilts, chair seats, etc. In standard dollhouse size. 48pp. 8¼ × 11.   23741-9 Pa. $1.95

THE DINOSAUR COLORING BOOK, Anthony Rao. 45 renderings of dinosaurs, fossil birds, turtles, other creatures of Mesozoic Era. Scientifically accurate. Captions. 48pp. 8¼ × 11.   24022-3 Pa. $2.50

JAPANESE DESIGN MOTIFS, Matsuya Co. Mon, or heraldic designs. Over 4000 typical, beautiful designs: birds, animals, flowers, swords, fans, geometrics; all beautifully stylized. 213pp. 11⅜ × 8¼. 22874-6 Pa. $7.95

THE TALE OF BENJAMIN BUNNY, Beatrix Potter. Peter Rabbit's cousin coaxes him back into Mr. McGregor's garden for a whole new set of adventures. All 27 full-color illustrations. 59pp. 4¼ × 5½. (Available in U.S. only) 21102-9 Pa. $1.75

THE TALE OF PETER RABBIT AND OTHER FAVORITE STORIES BOXED SET, Beatrix Potter. Seven of Beatrix Potter's best-loved tales including Peter Rabbit in a specially designed, durable boxed set. 4¼ × 5½. Total of 447pp. 158 color illustrations. (Available in U.S. only) 23903-9 Pa. $12.25

PRACTICAL MENTAL MAGIC, Theodore Annemann. Nearly 200 astonishing feats of mental magic revealed in step-by-step detail. Complete advice on staging, patter, etc. Illustrated. 320pp. 5⅜ × 8½. 24426-1 Pa. $5.95

CELEBRATED CASES OF JUDGE DEE (DEE GOONG AN), translated by Robert Van Gulik. Authentic 18th-century Chinese detective novel; Dee and associates solve three interlocked cases. Led to van Gulik's own stories with same characters. Extensive introduction. 9 illustrations. 237pp. 5⅜ × 8½. 23337-5 Pa. $4.95

CUT & FOLD EXTRATERRESTRIAL INVADERS THAT FLY, M. Grater. Stage your own lilliputian space battles. By following the step-by-step instructions and explanatory diagrams you can launch 22 full-color fliers into space. 36pp. 8¼ × 11. 24478-4 Pa. $2.95

CUT & ASSEMBLE VICTORIAN HOUSES, Edmund V. Gillon, Jr. Printed in full color on heavy cardboard stock, 4 authentic Victorian houses in H-O scale: Italian-style Villa, Octagon, Second Empire, Stick Style. 48pp. 9¼ × 12¼. 23849-0 Pa. $4.95

BEST SCIENCE FICTION STORIES OF H.G. WELLS, H.G. Wells. Full novel The Invisible Man, plus 17 short stories: "The Crystal Egg," "Aepyornis Island," "The Strange Orchid," etc. 303pp. 5⅜ × 8½. (Available in U.S. only) 21531-8 Pa. $4.95

TRADEMARK DESIGNS OF THE WORLD, Yusaku Kamekura. A lavish collection of nearly 700 trademarks, the work of Wright, Loewy, Klee, Binder, hundreds of others. 160pp. 8¾ × 8. (EJ) 24191-2 Pa. $5.95

THE ARTIST'S AND CRAFTSMAN'S GUIDE TO REDUCING, ENLARGING AND TRANSFERRING DESIGNS, Rita Weiss. Discover, reduce, enlarge, transfer designs from any objects to any craft project. 12pp. plus 16 sheets special graph paper. 8¼ × 11. 24142-4 Pa. $3.95

TREASURY OF JAPANESE DESIGNS AND MOTIFS FOR ARTISTS AND CRAFTSMEN, edited by Carol Belanger Grafton. Indispensable collection of 360 traditional Japanese designs and motifs redrawn in clean, crisp black-and-white, copyright-free illustrations. 96pp. 8¼ × 11. 24435-0 Pa. $4.50

CHANCERY CURSIVE STROKE BY STROKE, Arthur Baker. Instructions and illustrations for each stroke of each letter (upper and lower case) and numerals. 54 full-page plates. 64pp. 8¼ × 11. 24278-1 Pa. $2.50

THE ENJOYMENT AND USE OF COLOR, Walter Sargent. Color relationships, values, intensities; complementary colors, illumination, similar topics. Color in nature and art. 7 color plates, 29 illustrations. 274pp. 5⅜ × 8½. 20944-X Pa. $4.95

SCULPTURE PRINCIPLES AND PRACTICE, Louis Slobodkin. Step-by-step approach to clay, plaster, metals, stone; classical and modern. 253 drawings, photos. 255pp. 8⅛ × 11. 22960-2 Pa. $7.50

VICTORIAN FASHION PAPER DOLLS FROM HARPER'S BAZAR, 1867-1898, Theodore Menten. Four female dolls with 28 elegant high fashion costumes, printed in full color. 32pp. 9¼ × 12¼. 23453-3 Pa. $3.95

FLOPSY, MOPSY AND COTTONTAIL: A Little Book of Paper Dolls in Full Color, Susan LaBelle. Three dolls and 21 costumes (7 for each doll) show Peter Rabbit's siblings dressed for holidays, gardening, hiking, etc. Charming borders, captions. 48pp. 4¼ × 5½. (USCO) 24376-1 Pa. $2.50

NATIONAL LEAGUE BASEBALL CARD CLASSICS, Bert Randolph Sugar. 83 big-leaguers from 1909-69 on facsimile cards. Hubbell, Dean, Spahn, Brock plus advertising, info, no duplications. Perforated, detachable. 16pp. 8¼ × 11. 24308-7 Pa. $3.50

THE LOGICAL APPROACH TO CHESS, Dr. Max Euwe, et al. First-rate text of comprehensive strategy, tactics, theory for the amateur. No gambits to memorize, just a clear, logical approach. 224pp. 5⅜ × 8½. 24353-2 Pa. $4.50

MAGICK IN THEORY AND PRACTICE, Aleister Crowley. The summation of the thought and practice of the century's most famous necromancer, long hard to find. Crowley's best book. 436pp. 5⅜ × 8½. (Available in U.S. only) 23295-6 Pa. $6.95

THE HAUNTED HOTEL, Wilkie Collins. Collins' last great tale; doom and destiny in a Venetian palace. Praised by T.S. Eliot. 127pp. 5⅜ × 8½. 24333-8 Pa. $3.00

ART DECO DISPLAY ALPHABETS, Dan X. Solo. Wide variety of bold yet elegant lettering in handsome Art Deco styles. 100 complete fonts, with numerals, punctuation, more. 104pp. 8⅛ × 11. 24372-9 Pa. $4.50

CALLIGRAPHIC ALPHABETS, Arthur Baker. Nearly 150 complete alphabets by outstanding contemporary. Stimulating ideas; useful source for unique effects. 154 plates. 157pp. 8⅜ × 11¼. 21045-6 Pa. $5.95

ARTHUR BAKER'S HISTORIC CALLIGRAPHIC ALPHABETS, Arthur Baker. From monumental capitals of first-century Rome to humanistic cursive of 16th century, 33 alphabets in fresh interpretations. 88 plates. 96pp. 9 × 12. 24054-1 Pa. $4.50

LETTIE LANE PAPER DOLLS, Sheila Young. Genteel turn-of-the-century family very popular then and now. 24 paper dolls. 16 plates in full color. 32pp. 9¼ × 12¼. 24089-4 Pa. $3.95

KEYBOARD WORKS FOR SOLO INSTRUMENTS, G.F. Handel. 35 neglected works from Handel's vast oeuvre, originally jotted down as improvisations. Includes Eight Great Suites, others. New sequence. 174pp. 9⅜ × 12¼.
24338-9 Pa. $7.50

AMERICAN LEAGUE BASEBALL CARD CLASSICS, Bert Randolph Sugar. 82 stars from 1900s to 60s on facsimile cards. Ruth, Cobb, Mantle, Williams, plus advertising, info, no duplications. Perforated, detachable. 16pp. 8¼ × 11.
24286-2 Pa. $3.50

A TREASURY OF CHARTED DESIGNS FOR NEEDLEWORKERS, Georgia Gorham and Jeanne Warth. 141 charted designs: owl, cat with yarn, tulips, piano, spinning wheel, covered bridge, Victorian house and many others. 48pp. 8¼ × 11.
23558-0 Pa. $1.95

DANISH FLORAL CHARTED DESIGNS, Gerda Bengtsson. Exquisite collection of over 40 different florals: anemone, Iceland poppy, wild fruit, pansies, many others. 45 illustrations. 48pp. 8¼ × 11. 23957-8 Pa. $2.50

OLD PHILADELPHIA IN EARLY PHOTOGRAPHS 1839-1914, Robert F. Looney. 215 photographs: panoramas, street scenes, landmarks, President-elect Lincoln's visit, 1876 Centennial Exposition, much more. 230pp. 8⅞ × 11¾.
23345-6 Pa. $9.95

PRELUDE TO MATHEMATICS, W.W. Sawyer. Noted mathematician's lively, stimulating account of non-Euclidean geometry, matrices, determinants, group theory, other topics. Emphasis on novel, striking aspects. 224pp. 5⅜ × 8½.
24401-6 Pa. $4.50

ADVENTURES WITH A MICROSCOPE, Richard Headstrom. 59 adventures with clothing fibers, protozoa, ferns and lichens, roots and leaves, much more. 142 illustrations. 232pp. 5⅜ × 8½. 23471-1 Pa. $3.95

IDENTIFYING ANIMAL TRACKS: MAMMALS, BIRDS, AND OTHER ANIMALS OF THE EASTERN UNITED STATES, Richard Headstrom. For hunters, naturalists, scouts, nature-lovers. Diagrams of tracks, tips on identification. 128pp. 5⅜ × 8. 24442-3 Pa. $3.50

VICTORIAN FASHIONS AND COSTUMES FROM HARPER'S BAZAR, 1867-1898, edited by Stella Blum. Day costumes, evening wear, sports clothes, shoes, hats, other accessories in over 1,000 detailed engravings. 320pp. 9⅜ × 12¼.
22990-4 Pa. $10.95

EVERYDAY FASHIONS OF THE TWENTIES AS PICTURED IN SEARS AND OTHER CATALOGS, edited by Stella Blum. Actual dress of the Roaring Twenties, with text by Stella Blum. Over 750 illustrations, captions. 156pp. 9 × 12.
24134-3 Pa. $8.95

HALL OF FAME BASEBALL CARDS, edited by Bert Randolph Sugar. Cy Young, Ted Williams, Lou Gehrig, and many other Hall of Fame greats on 92 full-color, detachable reprints of early baseball cards. No duplication of cards with *Classic Baseball Cards*. 16pp. 8¼ × 11. 23624-2 Pa. $3.50

THE ART OF HAND LETTERING, Helm Wotzkow. Course in hand lettering, Roman, Gothic, Italic, Block, Script. Tools, proportions, optical aspects, individual variation. Very quality conscious. Hundreds of specimens. 320pp. 5⅜ × 8½.
21797-3 Pa. $5.95

HOW THE OTHER HALF LIVES, Jacob A. Riis. Journalistic record of filth, degradation, upward drive in New York immigrant slums, shops, around 1900. New edition includes 100 original Riis photos, monuments of early photography. 233pp. 10 × 7⅞. 22012-5 Pa. $9.95

CHINA AND ITS PEOPLE IN EARLY PHOTOGRAPHS, John Thomson. In 200 black-and-white photographs of exceptional quality photographic pioneer Thomson captures the mountains, dwellings, monuments and people of 19th-century China. 272pp. 9⅜ × 12¼. 24393-1 Pa. $13.95

GODEY COSTUME PLATES IN COLOR FOR DECOUPAGE AND FRAMING, edited by Eleanor Hasbrouk Rawlings. 24 full-color engravings depicting 19th-century Parisian haute couture. Printed on one side only. 56pp. 8¼ × 11. 23879-2 Pa. $3.95

ART NOUVEAU STAINED GLASS PATTERN BOOK, Ed Sibbett, Jr. 104 projects using well-known themes of Art Nouveau: swirling forms, florals, peacocks, and sensuous women. 60pp. 8¼ × 11. 23577-7 Pa. $3.95

QUICK AND EASY PATCHWORK ON THE SEWING MACHINE: Susan Aylsworth Murwin and Suzzy Payne. Instructions, diagrams show exactly how to machine sew 12 quilts. 48pp. of templates. 50 figures. 80pp. 8¼ × 11. 23770-2 Pa. $3.95

THE STANDARD BOOK OF QUILT MAKING AND COLLECTING, Marguerite Ickis. Full information, full-sized patterns for making 46 traditional quilts, also 150 other patterns. 483 illustrations. 273pp. 6⅞ × 9⅝. 20582-7 Pa. $5.95

LETTERING AND ALPHABETS, J. Albert Cavanagh. 85 complete alphabets lettered in various styles; instructions for spacing, roughs, brushwork. 121pp. 8¾ × 8. 20053-1 Pa. $3.95

LETTER FORMS: 110 COMPLETE ALPHABETS, Frederick Lambert. 110 sets of capital letters; 16 lower case alphabets; 70 sets of numbers and other symbols. 110pp. 8⅛ × 11. 22872-X Pa. $4.50

ORCHIDS AS HOUSE PLANTS, Rebecca Tyson Northen. Grow cattleyas and many other kinds of orchids—in a window, in a case, or under artificial light. 63 illustrations. 148pp. 5⅜ × 8½. 23261-1 Pa. $2.95

THE MUSHROOM HANDBOOK, Louis C.C. Krieger. Still the best popular handbook. Full descriptions of 259 species, extremely thorough text, poisons, folklore, etc. 32 color plates; 126 other illustrations. 560pp. 5⅜ × 8½. 21861-9 Pa. $8.50

THE DORÉ BIBLE ILLUSTRATIONS, Gustave Doré. All wonderful, detailed plates: Adam and Eve, Flood, Babylon, life of Jesus, etc. Brief King James text with each plate. 241 plates. 241pp. 9 × 12. 23004-X Pa. $8.95

THE BOOK OF KELLS: Selected Plates in Full Color, edited by Blanche Cirker. 32 full-page plates from greatest manuscript-icon of early Middle Ages. Fantastic, mysterious. Publisher's Note. Captions. 32pp. 9¾ × 12¼. 24345-1 Pa. $4.50

THE PERFECT WAGNERITE, George Bernard Shaw. Brilliant criticism of the Ring Cycle, with provocative interpretation of politics, economic theories behind the Ring. 136pp. 5⅜ × 8½. (EUK) 21707-8 Pa. $3.95

THE RIME OF THE ANCIENT MARINER, Gustave Doré, S.T. Coleridge. Doré's finest work, 34 plates capture moods, subtleties of poem. Full text. 77pp. 9¼ × 12. 22305-1 Pa. $4.95

SONGS OF INNOCENCE, William Blake. The first and most popular of Blake's famous "Illuminated Books," in a facsimile edition reproducing all 31 brightly colored plates. Additional printed text of each poem. 64pp. 5¼ × 7. 22764-2 Pa. $3.50

AN INTRODUCTION TO INFORMATION THEORY, J.R. Pierce. Second (1980) edition of most impressive non-technical account available. Encoding, entropy, noisy channel, related areas, etc. 320pp. 5⅜ × 8½. 24061-4 Pa. $5.95

THE DIVINE PROPORTION: A STUDY IN MATHEMATICAL BEAUTY, H.E. Huntley. "Divine proportion" or "golden ratio" in poetry, Pascal's triangle, philosophy, psychology, music, mathematical figures, etc. Excellent bridge between science and art. 58 figures. 185pp. 5⅜ × 8½. 22254-3 Pa. $4.50

THE DOVER NEW YORK WALKING GUIDE: From the Battery to Wall Street, Mary J. Shapiro. Superb inexpensive guide to historic buildings and locales in lower Manhattan: Trinity Church, Bowling Green, more. Complete Text; maps. 36 illustrations. 48pp. 3⅞ × 9¼. 24225-0 Pa. $2.50

NEW YORK THEN AND NOW, Edward B. Watson, Edmund V. Gillon, Jr. 83 important Manhattan sites: on facing pages early photographs (1875-1925) and 1976 photos by Gillon. 172 illustrations. 171pp. 9¼ × 10. 23361-8 Pa. $9.95

HISTORIC COSTUME IN PICTURES, Braun & Schneider. Over 1450 costumed figures from dawn of civilization to end of 19th century. English captions. 125 plates. 256pp. 8⅜ × 11¼. 23150-X Pa. $7.95

VICTORIAN AND EDWARDIAN FASHION: A Photographic Survey, Alison Gernsheim. First fashion history completely illustrated by contemporary photographs. Full text plus 235 photos, 1840-1914, in which many celebrities appear. 240pp. 6½ × 9¼. 24205-6 Pa. $6.00

CHARTED CHRISTMAS DESIGNS FOR COUNTED CROSS-STITCH AND OTHER NEEDLECRAFTS, Lindberg Press. Charted designs for 45 beautiful needlecraft projects with many yuletide and wintertime motifs. 48pp. 8¼ × 11. (EDNS) 24356-7 Pa. $2.50

101 FOLK DESIGNS FOR COUNTED CROSS-STITCH AND OTHER NEEDLE-CRAFTS, Carter Houck. 101 authentic charted folk designs in a wide array of lovely representations with many suggestions for effective use. 48pp. 8¼ × 11. 24369-9 Pa. $2.25

FIVE ACRES AND INDEPENDENCE, Maurice G. Kains. Great back-to-the-land classic explains basics of self-sufficient farming. The one book to get. 95 illustrations. 397pp. 5⅜ × 8½. 20974-1 Pa. $6.50

A MODERN HERBAL, Margaret Grieve. Much the fullest, most exact, most useful compilation of herbal material. Gigantic alphabetical encyclopedia, from aconite to zedoary, gives botanical information, medical properties, folklore, economic uses, and much else. Indispensable to serious reader. 161 illustrations. 888pp. 6½ × 9¼. (Available in U.S. only) 22798-7, 22799-5 Pa., Two-vol. set $17.00

DECORATIVE NAPKIN FOLDING FOR BEGINNERS, Lillian Oppenheimer and Natalie Epstein. 22 different napkin folds in the shape of a heart, clown's hat, love knot, etc. 63 drawings. 48pp. 8¼ × 11. 23797-4 Pa. $2.25

DECORATIVE LABELS FOR HOME CANNING, PRESERVING, AND OTHER HOUSEHOLD AND GIFT USES, Theodore Menten. 128 gummed, perforated labels, beautifully printed in 2 colors. 12 versions. Adhere to metal, glass, wood, ceramics. 24pp. 8¼ × 11. 23219-0 Pa. $3.50

EARLY AMERICAN STENCILS ON WALLS AND FURNITURE, Janet Waring. Thorough coverage of 19th-century folk art: techniques, artifacts, surviving specimens. 166 illustrations, 7 in color. 147pp. of text. 7⅞ × 10¾. 21906-2 Pa. $9.95

AMERICAN ANTIQUE WEATHERVANES, A.B. & W.T. Westervelt. Extensively illustrated 1883 catalog exhibiting over 550 copper weathervanes and finials. Excellent primary source by one of the principal manufacturers. 104pp. 6⅛ × 9¼. 24396-6 Pa. $3.95

ART STUDENTS' ANATOMY, Edmond J. Farris. Long favorite in art schools. Basic elements, common positions, actions. Full text, 158 illustrations. 159pp. 5⅜ × 8½. 20744-7 Pa. $3.95

BRIDGMAN'S LIFE DRAWING, George B. Bridgman. More than 500 drawings and text teach you to abstract the body into its major masses. Also specific areas of anatomy. 192pp. 6½ × 9¼. 22710-3 Pa. $4.50

COMPLETE PRELUDES AND ETUDES FOR SOLO PIANO, Frederic Chopin. All 26 Preludes, all 27 Etudes by greatest composer of piano music. Authoritative Paderewski edition. 224pp. 9 × 12. (Available in U.S. only) 24052-5 Pa. $7.50

PIANO MUSIC 1888-1905, Claude Debussy. Deux Arabesques, Suite Bergamesque, Masques, 1st series of Images, etc. 9 others, in corrected editions. 175pp. 9⅜ × 12¼. 22771-5 Pa. $6.95

TEDDY BEAR IRON-ON TRANSFER PATTERNS, Ted Menten. 80 iron-on transfer patterns of male and female Teddys in a wide variety of activities, poses, sizes. 48pp. 8¼ × 11. 24596-9 Pa. $2.25

A PICTURE HISTORY OF THE BROOKLYN BRIDGE, M.J. Shapiro. Profusely illustrated account of greatest engineering achievement of 19th century. 167 rare photos & engravings recall construction, human drama. Extensive, detailed text. 122pp. 8¼ × 11. 24403-2 Pa. $7.95

NEW YORK IN THE THIRTIES, Berenice Abbott. Noted photographer's fascinating study shows new buildings that have become famous and old sights that have disappeared forever. 97 photographs. 97pp. 11⅜ × 10. 22967-X Pa. $7.50

MATHEMATICAL TABLES AND FORMULAS, Robert D. Carmichael and Edwin R. Smith. Logarithms, sines, tangents, trig functions, powers, roots, reciprocals, exponential and hyperbolic functions, formulas and theorems. 269pp. 5⅜ × 8½. 60111-0 Pa. $4.95

HANDBOOK OF MATHEMATICAL FUNCTIONS WITH FORMULAS, GRAPHS, AND MATHEMATICAL TABLES, edited by Milton Abramowitz and Irene A. Stegun. Vast compendium: 29 sets of tables, some to as high as 20 places. 1,046pp. 8 × 10½. 61272-4 Pa. $21.95

REASON IN ART, George Santayana. Renowned philosopher's provocative, seminal treatment of basis of art in instinct and experience. Volume Four of *The Life of Reason*. 230pp. 5⅜ × 8. 24358-3 Pa. $4.50

LANGUAGE, TRUTH AND LOGIC, Alfred J. Ayer. Famous, clear introduction to Vienna, Cambridge schools of Logical Positivism. Role of philosophy, elimination of metaphysics, nature of analysis, etc. 160pp. 5⅜ × 8½. (USCO) 20010-8 Pa. $2.95

BASIC ELECTRONICS, U.S. Bureau of Naval Personnel. Electron tubes, circuits, antennas, AM, FM, and CW transmission and receiving, etc. 560 illustrations. 567pp. 6½ × 9¼. 21076-6 Pa. $9.95

THE ART DECO STYLE, edited by Theodore Menten. Furniture, jewelry, metalwork, ceramics, fabrics, lighting fixtures, interior decors, exteriors, graphics from pure French sources. Over 400 photographs. 183pp. 8⅜ × 11¼. 22824-X Pa. $7.95

THE FOUR BOOKS OF ARCHITECTURE, Andrea Palladio. 16th-century classic covers classical architectural remains, Renaissance revivals, classical orders, etc. 1738 Ware English edition. 216 plates. 110pp. of text. 9½ × 12¾. 21308-0 Pa. $11.95

THE WIT AND HUMOR OF OSCAR WILDE, edited by Alvin Redman. More than 1000 ripostes, paradoxes, wisecracks: Work is the curse of the drinking classes, I can resist everything except temptations, etc. 258pp. 5⅜ × 8½. 20602-5 Pa. $4.50

THE DEVIL'S DICTIONARY, Ambrose Bierce. Barbed, bitter, brilliant witticisms in the form of a dictionary. Best, most ferocious satire America has produced. 145pp. 5⅜ × 8½. 20487-1 Pa. $2.95

ERTÉ'S FASHION DESIGNS, Erté. 210 black-and-white inventions from *Harper's Bazar*, 1918-32, plus 8pp. full-color covers. Captions. 88pp. 9 × 12. 24203-X Pa. $7.95

ERTÉ GRAPHICS, Erté. Collection of striking color graphics: *Seasons, Alphabet, Numerals, Aces* and *Precious Stones*. 50 plates, including 4 on covers. 48pp. 9⅜ × 12¼. 23580-7 Pa. $6.95

PAPER FOLDING FOR BEGINNERS, William D. Murray and Francis J. Rigney. Clearest book for making origami sail boats, roosters, frogs that move legs, etc. 40 projects. More than 275 illustrations. 94pp. 5⅜ × 8½. 20713-7 Pa. $2.50

ORIGAMI FOR THE ENTHUSIAST, John Montroll. Fish, ostrich, peacock, squirrel, rhinoceros, Pegasus, 19 other intricate subjects. Instructions. Diagrams. 128pp. 9 × 12. 23799-0 Pa. $5.95

CROCHETING NOVELTY POT HOLDERS, edited by Linda Macho. 64 useful, whimsical pot holders feature kitchen themes, animals, flowers, other novelties. Surprisingly easy to crochet. Complete instructions. 48pp. 8¼ × 11. 24296-X Pa. $1.95

CROCHETING DOILIES, edited by Rita Weiss. Irish Crochet, Jewel, Star Wheel, Vanity Fair and more. Also luncheon and console sets, runners and centerpieces. 51 illustrations. 48pp. 8¼ × 11. 23424-X Pa. $2.75

YUCATAN BEFORE AND AFTER THE CONQUEST, Diego de Landa. Only significant account of Yucatan written in the early post-Conquest era. Translated by William Gates. Over 120 illustrations. 162pp. 5⅜ × 8½.   23622-6 Pa. $3.95

ORNATE PICTORIAL CALLIGRAPHY, E.A. Lupfer. Complete instructions, over 150 examples help you create magnificent "flourishes" from which beautiful animals and objects gracefully emerge. 8⅛ × 11.   21957-7 Pa. $3.50

DOLLY DINGLE PAPER DOLLS, Grace Drayton. Cute chubby children by same artist who did Campbell Kids. Rare plates from 1910s. 30 paper dolls and over 100 outfits reproduced in full color. 32pp. 9¼ × 12¼.   23711-7 Pa. $3.50

CURIOUS GEORGE PAPER DOLLS IN FULL COLOR, H. A. Rey, Kathy Allert. Naughty little monkey-hero of children's books in two doll figures, plus 48 full-color costumes: pirate, Indian chief, fireman, more. 32pp. 9¼ × 12¼.
24386-9 Pa. $3.50

GERMAN: HOW TO SPEAK AND WRITE IT, Joseph Rosenberg. Like *French, How to Speak and Write It.* Very rich modern course, with a wealth of pictorial material. 330 illustrations. 384pp. 5⅜ × 8½.   20271-2 Pa. $4.95

CATS AND KITTENS: 24 Ready-to-Mail Color Photo Postcards, D. Holby. Handsome collection; feline in a variety of adorable poses. Identifications. 12pp. on postcard stock. 8¼ × 11.   24469-5 Pa. $2.95

MARILYN MONROE PAPER DOLLS, Tom Tierney. 31 full-color designs on heavy stock, from *The Asphalt Jungle,Gentlemen Prefer Blondes,* 22 others.1 doll. 16 plates. 32pp. 9⅜ × 12¼.   23769-9 Pa. $3.95

FUNDAMENTALS OF LAYOUT, F.H. Wills. All phases of layout design discussed and illustrated in 121 illustrations. Indispensable as student's text or handbook for professional. 124pp. 8⅛.× 11.   21279-3 Pa. $4.50

FANTASTIC SUPER STICKERS, Ed Sibbett, Jr. 75 colorful pressure-sensitive stickers. Peel off and place for a touch of pizzazz: clowns, penguins, teddy bears, etc. Full color. 16pp. 8¼ × 11.   24471-7 Pa. $3.50

LABELS FOR ALL OCCASIONS, Ed Sibbett, Jr. 6 labels each of 16 different designs—baroque, art nouveau, art deco, Pennsylvania Dutch, etc.—in full color. 24pp. 8¼ × 11.   23688-9 Pa. $3.95

HOW TO CALCULATE QUICKLY: RAPID METHODS IN BASIC MATHE-MATICS, Henry Sticker. Addition, subtraction, multiplication, division, checks, etc. More than 8000 problems, solutions. 185pp. 5 × 7¼.   20295-X Pa. $2.95

THE CAT COLORING BOOK, Karen Baldauski. Handsome, realistic renderings of 40 splendid felines, from American shorthair to exotic types. 44 plates. Captions. 48pp. 8¼ × 11.   24011-8 Pa. $2.50

THE TALE OF PETER RABBIT, Beatrix Potter. The inimitable Peter's terrifying adventure in Mr. McGregor's garden, with all 27 wonderful, full-color Potter illustrations. 55pp. 4¼ × 5½. (Available in U.S. only)   22827-4 Pa. $1.75

BASIC ELECTRICITY, U.S. Bureau of Naval Personnel. Batteries, circuits, conductors, AC and DC, inductance and capacitance, generators, motors, trans-formers, amplifiers, etc. 349 illustrations. 448pp. 6½ × 9¼.   20973-3 Pa. $7.95

SOURCE BOOK OF MEDICAL HISTORY, edited by Logan Clendening, M.D. Original accounts ranging from Ancient Egypt and Greece to discovery of X-rays: Galen, Pasteur, Lavoisier, Harvey, Parkinson, others. 685pp. 5⅜ × 8½.
20621-1 Pa. $11.95

THE ROSE AND THE KEY, J.S. Lefanu. Superb mystery novel from Irish master. Dark doings among an ancient and aristocratic English family. Well-drawn characters; capital suspense. Introduction by N. Donaldson. 448pp. 5⅜ × 8½.
24377-X Pa. $6.95

SOUTH WIND, Norman Douglas. Witty, elegant novel of ideas set on languorous Meditterranean island of Nepenthe. Elegant prose, glittering epigrams, mordant satire. 1917 masterpiece. 416pp. 5⅜ × 8½. (Available in U.S. only)
24361-3 Pa. $5.95

RUSSELL'S CIVIL WAR PHOTOGRAPHS, Capt. A.J. Russell. 116 rare Civil War Photos: Bull Run, Virginia campaigns, bridges, railroads, Richmond, Lincoln's funeral car. Many never seen before. Captions. 128pp. 9⅜ × 12¼.
24283-8 Pa. $7.95

PHOTOGRAPHS BY MAN RAY: 105 Works, 1920-1934. Nudes, still lifes, landscapes, women's faces, celebrity portraits (Dali, Matisse, Picasso, others), rayographs. Reprinted from rare gravure edition. 128pp. 9⅜ × 12¼.
23842-3 Pa. $8.95

STAR NAMES: THEIR LORE AND MEANING, Richard H. Allen. Star names, the zodiac, constellations: folklore and literature associated with heavens. The basic book of its field, fascinating reading. 563pp. 5⅜ × 8½.   21079-0 Pa. $7.95

BURNHAM'S CELESTIAL HANDBOOK, Robert Burnham, Jr. Thorough guide to the stars beyond our solar system. Exhaustive treatment. Alphabetical by constellation: Andromeda to Cetus in Vol. 1; Chamaeleon to Orion in Vol. 2; and Pavo to Vulpecula in Vol. 3. Hundreds of illustrations. Index in Vol. 3. 2000pp. 6⅛ × 9¼.   23567-X, 23568-8, 23673-0 Pa. Three-vol. set $37.85

THE ART NOUVEAU STYLE BOOK OF ALPHONSE MUCHA, Alphonse Mucha. All 72 plates from *Documents Decoratifs* in original color. Stunning, essential work of Art Nouveau. 80pp. 9⅜ × 12¼.   24044-4 Pa. $8.95

DESIGNS BY ERTE; FASHION DRAWINGS AND ILLUSTRATIONS FROM "HARPER'S BAZAR," Erte. 310 fabulous line drawings and 14 *Harper's Bazar* covers, 8 in. full color. Erte's exotic temptresses with tassels, fur muffs, long trains, coifs, more. 129pp. 9⅜ × 12¼.   23397-9 Pa. $8.95

HISTORY OF STRENGTH OF MATERIALS, Stephen P. Timoshenko. Excellent historical survey of the strength of materials with many references to the theories of elasticity and structure. 245 figures. 452pp. 5⅜ × 8½. 61187-6 Pa. $9.95

Prices subject to change without notice.
Available at your book dealer or write for free catalog to Dept. GI, Dover Publications, Inc., 31 East 2nd St. Mineola, N.Y. 11501. Dover publishes more than 175 books each year on science, elementary and advanced mathematics, biology, music, art, literary history, social sciences and other areas.